Lecture Series on

MAGNETOSTATICS &

ELECTRODYNAMICS

Magnetic Effects of Steady Current :

Lorentz force:

The force exerted on a point charge 'q' by an electrostatic field $\vec{E}$ is : $\vec{Fe} = q\vec{E}$. In our discussion of this force, we assumed the charge q was at rest. If we allow the charge q to move with a velocity $\vec{v}$, an additional force may appear. This force is called magnetic force or the Lorentz force, acting on 'q' is given by (SI units):

$$\vec{F}_m = q(\vec{v} \times \vec{B})$$

where the quantity $\vec{B}$ is known as "magnetic induction" or the magnetic field vector.

The Lorentz force is perpendicular to the plane containing the vectors $\vec{v}$ & $\vec{B}$. If 'q' is +ve, the direction of the force vector $\vec{F}_m$ is obtained by rotating a right handed cock screw from $\vec{v}$ to $\vec{B}$; the direction of advance of the tip of this screw gives the direction of $\vec{F}_m$ is:-

$$\vec{F}_m = q(\vec{v} \times \vec{B}) \Rightarrow |\vec{F}_m| = Bqu\sin\theta$$

$\theta \rightarrow$ angle between $\vec{v} \times \vec{B}$.

Work done by $\vec{F}_m$:-

$$Dw = \vec{F}_m . d\vec{x} = \vec{F}_m . \vec{v}\, dt$$

$$= q(\vec{v} \times \vec{B}).\vec{v}\, dt = 0$$

Thus, the Lorentz force acting alone cannot do any work on the charged particles.

If both the electrostatic field $\vec{E}$ and the magnetic induction $\vec{B}$ are present, the total force, known as the electromagnetic force (the electromagnetic force, and not the magnetic force, is referred to as Lorentz force by some authors) acting on a point charge 'q' is given by:-

$$\vec{F} = q(\vec{E} \times \vec{v} \times \vec{B})$$

This force governs the motion of the charged particles in electromagnetic. The behaviour of the charged particles in electromagnetic fields is the subject matter of electrodynamics. If the equations of motion are these of classical physics, the topic is referred to as "classical electrodynamics". If the equations are of quantum physics, the subject is known as "quantum electrodynamics".

SI unit of B:-

$$[B] = [\frac{volt \times second}{m^2}]$$

The quantity volt-second is called Weber (Wb). Thus, the SI unit of B is Wb/m^2 or tesla (T).

Force on a current carrying conductor

Consider an element $\vec{dl}$ of a conductor carrying a current I. The direction of $\vec{dl}$ is same as that of I, so that $\vec{dl}$ is parallel to the velocity $\vec{v}$ of the charge carriers inside the conductor (The quantity $I\vec{dl}$ is known as current element).

∴ The no. of charge carriers inside the element $\vec{dl} = Na\vec{dl}$

N → No. of charges / vol.

A → Area of cross section of the conductor.

Each of these carriers experiences a magnetic force $d\vec{F} = NAdlq\,\vec{v} \times \vec{B}$

In equilibrium, this force is transmitted to the atoms of the conductor, the charges being unable to escape from the wire. ∴ the last equations gives the force acting on the conductor itself.

If several types of charge carriers are present, a summation is to be included.

∴ $\vec{dl}$ is parallel to $\vec{v}$, we have:-

$$d\vec{F} = NqvA\,\vec{dl} \times \vec{B} = I(\vec{dl} \times \vec{B})$$

where I → current in the elemental $dl = NqvA$

For a finite length of the conductor, the force F exerted on it is obtained by integrations;

$$\vec{F} = I\int (\vec{dl} \times \vec{B})$$

where I → current in the elemental $dl = NqvA$

For a finite length of the conductor, the force F exerted on it is obtained by integrations ;

$$\vec{F} = I\int \vec{dl} \times \vec{B}$$

If $\vec{B}$ is uniform, i.e., independent of position co-ordinates, we have,

$$\vec{F} = I(\int f\,\vec{dl}) \times \vec{B}$$

when a closed loop is considered, $\oint \vec{dl} = 0$. $\therefore$ force on a closed current loop is zero when $\vec{B}$ is uniform.

Origin of B

So far, the sources of magnetic field B have not been discussed. Experimental studies show that the origin of the magnetostatics field is steady electric current. In fact magnetic field produced by a permanent magnet can also be explained by representing the magnet as a current distribution.

We know that the electrostatic field $\vec{E}$ satisfies the relation :

$$\vec{\nabla}.\vec{B} = \rho/\varepsilon_o \text{ or } \vec{\nabla}.\vec{\nabla} = \rho$$

where $\rho \rightarrow$ electrostatic charge density.

This means we have source for $\vec{E}$ or $\vec{D}$, from which the field diverges.

In analogy, we can write for the magnetic field :

$$\vec{\nabla}.\vec{B} = Km\rho m;$$

$K_m \rightarrow$ Constant depending on the choice of units.

$\rho_m \rightarrow$ Magnetic charge density.

However, there is no experimental evidence as yet for the magnetic monopoles or charges. Free isolated magnetic poles do not exist. Magnetic poles are equal as opposite and occur in pairs. $\therefore$ ρ_m must be zero. Hence the static magnetic field satisfies the relationship :-

$$\vec{\nabla}.\vec{B} = 0$$

This means we do not have a source for $\vec{B}$ s hence there is no divergence.

In addition, experiments show that;

$\vec{\nabla} \times \vec{B} = \mu_0 \vec{j}$ (unlike electrostatic field, where $\vec{\nabla} \times \vec{E} = 0$. This means magnetic field curls around a paint in question. Hence, the $\vec{B}$ field are always in a loop, with no source).

where $\vec{j} \rightarrow$ conduction current density (A/m^2).

$\mu_0 \rightarrow$ Permeability of free space $= 4\pi \times 10^{-7} H/m$.

The equation $\vec{\Delta} \times \vec{B} = \mu_0 \vec{j}$ is known as Amperes circuital law for static in point form (or differential form).

Check :-

$$\vec{\nabla} \times \vec{B} = \mu_0 \vec{j}$$

or,

$$\vec{\nabla} . (\vec{\nabla} \times \vec{B}) = \mu_0 (\vec{\nabla} . \vec{j})$$

But $\vec{\nabla} . (\vec{\nabla} \times \vec{B}) = 0 \Rightarrow \vec{\nabla} . \vec{j} = 0$

But from the equation of continuity;

$$\vec{\nabla} . \vec{j} + \frac{\partial \rho}{\partial t} = 0$$

$$\Rightarrow \frac{\partial \rho}{\partial t} = 0$$

Thus, the charge density ρ must be independent of time, which no the condition of steady current. The Ampere's circuital law will be modified if the current changes with time (i.e., $\partial \rho / \partial \neq 0$, non-steady state or current).

From $\vec{\nabla} . \times \vec{B} = 0$

Integrating oner a volume V;

$$\int_v (\vec{\nabla} . \vec{B}) dv = 0 \Rightarrow \oint_s \vec{B} . d\vec{S} = 0 \dots\dots\text{by divergence theorem.}$$

$S \rightarrow$ surface bounding the Vol. V.

The quantity ; $\phi = \int_s \int_{s_1} \vec{B} . d\vec{S}$ is known as magnetic flux through the surface S_1s is measured in Webers (Wb) in SI units.

The equations $\oint_s \vec{B} . d\vec{S}$ shows that the magnetic flux through any closed surface S is zero. In other words, the lines of $\vec{B}$ are continuous or forms loops. There is no magnetic charge or monopole or which these lines originate or terminate.

Ampere's Circuital Law (In Integral Forms)

We have $\vec{\nabla} \times \vec{B} = \mu_0 \vec{j}$

Surface integrating on both sides;

$$\int_s (\vec{\nabla} \times \vec{B}).d\vec{s} = \mu_0 \int_s \vec{j}.d\vec{s} = \mu_0 I, \text{ where } I = \int_s \vec{j}.d\vec{s}$$

By Stokes theorem;

$$\int_s \vec{B}.\vec{dl} = \mu_0 I \rightarrow \text{Ampere's Circuital law in integral form.}$$

Ampere's circuital law (in SI units) : The line integral form.

Like Gauss' law, Ampere's law is always true (for steady currents), but it is not always useful. Only when the symmetry of the problem enables one to pull $\vec{B}$ outside the integral $\oint \vec{B}.\vec{dl}$ can one calculate the magnetic field from Ampere's law.

Just as Gaussian surface, we define "Amperian loop", where we get the symmetry.

1) The direction of $\vec{B}$ s an infinitesimal length $\vec{dl}$ of the Amperian loop should be such that $\oint \vec{B}.\vec{dl} = \phi_l Bdl$, i.e., the angle should be $0°$.

2) B should be uniform s constant throughout such that $\phi_l Bdl = \phi_l.dl$.

The standard current configurations which can be handled by Ampere's law are :-

1) Infinite straight lines.
2) Infinite planes.
3) Infinite solenoids.
4) Toroids.

Applications of Amperes' Circuital Law:-

Magnetic field due to a long straight current carrying conductor; consider an infinitely long straight wire carrying a current I. The wire is assumed to be infinitely long. The radius of the wire is 'a'. We have to find $\vec{B}$ at a distinction of 'r' from the wire, such that $r \geq a$ and also for $r<a$.

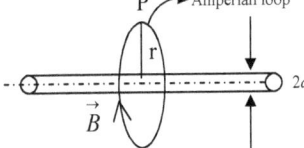

First, we find the direction of $\vec{B}$. If one grabs the wire with one's right hand-thumb in the direction of the current – fingers curb around in the direction of the magnetic field. Thus, $\vec{B}$ is circumferential, circling around the wire as indicated. We are to find $\vec{B}$ at P as shown :-

We defined an Amperian lop of radius r. Obviously $\vec{B}$ and $\vec{dl}$ (a small element of the loop) one $0°$ to each other.

$$\therefore \oint_l \vec{B}.\vec{dl} = \oint_l B\,dl$$

Also, as the wire is infinitely long, the end effects are neglected and hence $\vec{B}$ is constant and uniform.

$$\therefore \oint_l B\,dl = B\oint_l dl$$

$\therefore$ By Ampere's circuital law;

$B\oint_l dl = \mu_0 I_{in} = \mu_0 I$ (I_{in}=I, current enclosed by the Amperian loop)

M, $B(2\pi r) = \mu_0 I \Rightarrow B = \dfrac{\mu_0 I}{2\pi r} \Rightarrow \vec{B} = \dfrac{\mu_0 I}{2\pi r}\hat{e\phi}$

If the point P is inside the wire, such that $r < a$, we approach in the similar way, by defining on Amperian loop of radius r through ρ and observing that $\oint_l \vec{B}.\vec{dl} = B\oint_l dl$.

By Amperes' law, $\oint_l \vec{B}.\vec{dl} = B\oint_l dl = \mu_0 I_{in}$.

$I_{in} \rightarrow$ current enclosed by the Amperian loop.

$I_{in} = \pi r^2 j.$ But $j = \dfrac{I}{\pi a^2}$

$\therefore I_{in} = \dfrac{\pi r^2 I}{\pi a^2} = \dfrac{r^2}{a^2}I$

$\therefore B\oint_l dl = \mu_0 \dfrac{r^2}{a^2}I \Rightarrow B(2\pi r) = \mu_0 \dfrac{r^2}{a^2}I \Rightarrow \vec{B} = \dfrac{\mu_0 Ir}{2\pi a^2}\hat{e\phi}$

$\therefore \vec{B} \begin{cases} \dfrac{\mu_0 I_r}{2\pi a^2}\hat{e\phi} \; ; \; r < a \\[2mm] \dfrac{\mu_0 I_r}{2\pi r}\hat{e\phi} \; ; \; r \ge a \end{cases}$

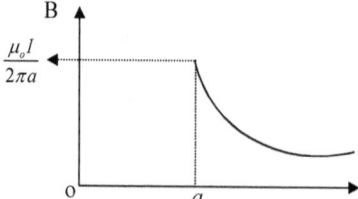

Magnetic field due to an infinitely long solenoid, consisting of 'n' closely wound turns per unit length on a cylinder of radius 'R' and carrying a steady current I.

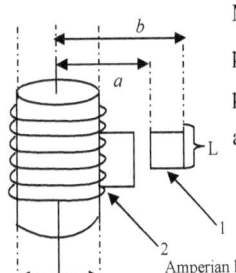

Magnetic field of an infinite, closely would solenoid points parallel to the axis. From the right hand rule, we expert that it points upwards inside the solenoid and downwards outside. It approaches zero as we go for away.

Amperian loops

Applying Amperes' law at the Amperian loop 1 as shown;

$$\oint_l \vec{B}.dl = [B(a) - B(b)]L = \mu_0 I_{in} = 0 \text{ (as the current enclosed by the Amperian loop is zero)}.$$

At the loop 2, we have;

$$\oint_l \vec{B}.\vec{dl} = \mu_0 I_{in} = BL = \mu_0 IN$$

where B→ field inside the selenoid. The right side of the loop contributes nothing; ∴B=0 outside. Thus;

$$\vec{B} = \begin{cases} \mu_0 WI \hat{k}; insi\det hesolenoid; n = \dfrac{N}{L} \\ 0; outside \ the \ solenoid. \end{cases}$$

Force between infinitely long parallel current carrying conductors

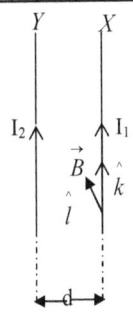

Let I_1 and I_2 be the currents in the two wires X and Y, spaced at a distance of 'd' apart in free space. The $\vec{B}$ field produced by the current I_2 of Y at any point in X is given by : $B = \dfrac{\mu_0 I_2}{2\pi d}$, where the radii of the wires are much smaller than 'd'. The direction of $\vec{B}$ is at right angles to the plane of the wire. The force per unit length exerted or the wire X is :-

$$\vec{F}_0 = I_1 k \times \vec{B}$$

or, $\vec{F}_0 = \dfrac{\mu_0 I_1 I_2}{2\pi d} (\hat{k} \times \hat{i}) N.$

where, $\hat{i}$ is a unit vector in the direction of $\vec{B}$. The force $\vec{F_0}$ is directed towards the wire Y, if I_1 is I_2 flow in the same direction. On the other hand, F_0 is directed away from the wire Y, if I_1 and I_2 flow is the opposite anti-parallel currents repel each other.

The magnitude of the force per metre is :

$$F_0 = \frac{\mu_0 I_1 I_2}{2\pi d} \, N/m$$

Magnetic Vector Potential

Just as $\vec{\nabla} \times \vec{E} = 0$ permitted us to introduce a scalor potential (v) in electrostatics, such that $\vec{E} = -\vec{\nabla} V$; so, $\vec{\nabla}.\vec{B} = 0$ innites the introduction of a vector potential $\vec{A}$ in magnetostatics :-

As $\vec{\nabla}.\vec{B} = 0 \Rightarrow \vec{B} = \vec{\nabla} \times \vec{A}$

where $\vec{A} \rightarrow$ magnetic vector potential (SI $\rightarrow$ wb/m).

But, we know that,

$$\vec{\nabla} \times \vec{B} = \mu_0 \vec{j}$$
$$\therefore \vec{\nabla} \times (\vec{\nabla} \times \vec{A}) = \mu_0 \vec{j}$$

For static case, we assume; $\vec{\nabla}.\vec{A} = 0$

$$\therefore \vec{\nabla}^2 \vec{A} = \mu_0 \vec{j}$$

The equations for the Cartesian components of $\vec{A}$ are;

$$\nabla^2 A_x = -\mu_0 j_x; \nabla^2 A_y = -\mu_0 j_y; \nabla^2 A_z = -\mu_0 j_z$$

Each of these component equations has the form of Poisson's equations $\nabla^2 v = -\rho/\varepsilon_0$, in electrostatics, which has the solution.

$$\nabla(\vec{r}) = \frac{1}{4\pi\varepsilon_o} \int \frac{\rho(\vec{r_0})dv_0}{|\vec{r}-\vec{r_0}|}$$

Using this as a guide, the solution for A_i, where i stands for x, y & z is written as;

$$A_i(\vec{r}) = \frac{\mu_0}{4\pi} \int \frac{j(\vec{r_0})dv_0}{|\vec{r}-\vec{r_0}|}$$

Thus, we can write;

$$\vec{A}(r) = \frac{\mu_0}{4\pi} \int \frac{\vec{j}(\vec{r_0}) dv_0}{l}$$

where $l = |\vec{r} - \vec{r_0}|$

For a current element I flowing in wire, $\vec{j}\,d\vec{v_0}$ is replaced by $I\,\vec{dl_0}$, where $\vec{dl_0}$ is a differential length along the wire. Thus, for a line current, we get:-

$$\vec{A}(r) = \frac{\mu_0}{4\pi} \int \frac{I(\vec{r_0}) dl_0}{l}$$

The Best-Savants' Law

For an elemental current Idl_0, the magnetic induction at a distinction 'l' from it is:-

$$d\vec{B}(r) = \frac{\mu_0}{4\pi} \int . \frac{I\,\vec{dl_0} \times l}{l^3}$$

or, $$\vec{B}(r) = \frac{\mu_0}{4\pi} \int . \frac{I\,\vec{dl_0} \times \vec{l}}{l^3}$$

The above two equations are known as Best-Savart's law. These are helpful in calculating $\vec{B}$ just as coulomb's law in calculating $\vec{E}$ in electrostatics. The above two equations are also known as Laplace's formulas expressing the magnetic field due to a current. The last equations states that the magnetic field dB at a point P due to a small element 'ab' of length dl_0 on a wire carrying a current I is :-

(i)   directly proportional to the current I.

(ii) Directly proportional to dlo and to sinθ, where θ is the angle between the vector $\vec{dl_0}$ and the vector $\vec{ab}$, represented by $\vec{l_0}$.

(iii) Inversely proportional to l^2.

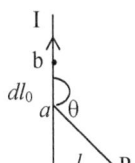

The direction of the field $\vec{dB}$ is $\perp^r$ to the plane containing the wire and the point P. The equation states that the total field at P due to the wire is given by the vector sum of the fields due to each small element, such as $ab = dl_0$.

Applications of Biot-Savarto' Law

(1) Magnetic induction due to the current in a straight wire of finite length.

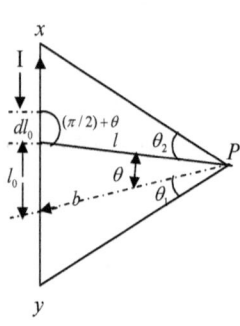

Let XY be a straight wire segment carrying a current I and P be appoint distant D from XY. The magnetic field at P is to be determined. Biot-Savart (or Laplace's) formula shows that the magnetic field at P due to an element dl_0 of the wire is;

$$dB = \frac{\mu_0}{4\pi}\frac{Idl_0}{l^2}\sin(\frac{\pi}{2}+\theta)$$

where $l \rightarrow$ distinction of the element dl_0 from P and θ as indicated

as $l_0 = D\tan\theta$

$$\therefore dl_0 = D\sec^2 d\theta$$

also $l = D\sec\theta$

$\therefore$ we obtain;

$$dB = \frac{\mu_0 I}{4\pi D}\cos\theta d\theta$$

$\therefore$ The total field at P due to the entire wire is ;

$$B = \frac{\mu_0 I}{4\pi D}\int_{-\theta_1}^{\theta_2}\cos\theta = \frac{\mu_0 I}{4\pi D}(\sin\theta_2 + \sin\theta_1)$$

For an infinitely long wire, $\theta_1 = \theta_1 = \pi/2$; so that B$=\frac{\mu_0 I}{2\pi D}$; which with D=r agrees

with the result obtained from Ampere's Circuital law. The direction of $\vec{B}$ is $\perp'$ to the plane of the paper and into it.

(2)   Magnetic field at a point or the axis of a circular conductor carrying current. Let P be a point on the axis of a circular loop carrying a current I is to be determined. Consider an element of length dl_0 on the loop. Biot-Savart's formula shows that the field at P due to the element $\vec{dl_0}$ is:-

$$dB = \frac{\mu_0 I}{4\pi}\frac{\vec{dl_0}\times\vec{l}}{l^3}, \text{where } l \rightarrow \text{ distinction of P from the element } dl_0. \ \therefore \text{ the angle}$$

between $\vec{dl_0}$ & $\vec{l}$ is $\pi/2$; we have :-

$$dB = \frac{\mu_0 Idl_0}{4\pi l^2}$$

This field acts perpendicularly to the plane containing $\vec{dl_0}$ & $\vec{l}$. Set this field be represented by the vector $\vec{PT}$. This vector is resolved into two components. $\vec{PV}$ along the axis of the loops $\vec{PH}$ perpendicular to the axis. Because of summetry, when the entire loop is considered, the components perpendicular to the axis cancel out and only the components along the axis contribute.

New, $PV = PT \sin \angle HPT = PT\left(\dfrac{a}{l}\right)$; where $a \rightarrow$ radius of the loop. Hence the component of $\vec{dB}$ along the loop axis is

$$\frac{\mu_0 I}{4\pi} \cdot \frac{dl_0}{l^2}\left(\frac{a}{l}\right) = \frac{\mu_0 Ia}{4\pi l^3} dl_0$$

The magnetic field due to the entire loop is :-

$$B = \frac{\mu_0 Ia}{4\pi l^3} \int dl_0 = \frac{\mu_0 Ia}{4\pi l^3}(2\pi a) = \frac{\mu_0 Ia^2}{2l^3}$$

or, $\vec{B} = \dfrac{\mu_0 Ia^2}{2(a^2 + z^2)^{3/2}}\hat{k}$

where $z \rightarrow$ distinction of P from 0.

If the coil has N number of closely wound terms, then $\int dl_0 = 2\pi Na$.

Hence, $\vec{B} = \dfrac{\mu_0 INa^2}{2(a^2 + z^2)^{3/2}}\hat{k}$

$\vec{B}$ at the centre of the loop is obtained by putting z=0, $\therefore$ we get:-

$\vec{B} = \dfrac{\mu_0 I}{2a}\hat{k}$; at the centre of the loop (z=0).

(3)   Magnetic induction at a point on the axis of a current carrying arc.

The current carrying arc is shown. We have to find $\vec{B}$ at P. By Biot-Savart's law;

$$d\vec{B} = \frac{\mu I}{4\pi}\frac{\vec{dl_0} \times \vec{r}}{r^3} = \frac{\mu_0 I}{4\pi}\frac{dl_0}{a^2}$$

$$\therefore B = \int dB = -\frac{\mu_0 I}{4\pi a^2}\int dl_0 = \frac{\mu_0 I}{4Ta}\theta; \quad \theta \rightarrow \text{in radians.}$$

The directions of $\vec{B}$ is $\perp'$ to the surface, pointing outwards.

(4) Axial magnetic field of a solenoid :

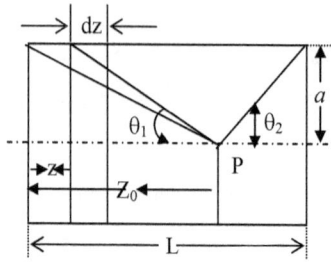

A solenoid is a device containing a number of twins would uniformly on a cylindrical form. Set N be the turns, L be the length and 'a' be the radius of the solenoid carrying a current I. The magnetic induction at a point P on the axis of the solenoid is to be determined. Set Z_0 be the distinction of P from the left end of the solenoid. We divide the length L into elements of width dz. One such element is shown in fig. Using the results of the earlier derivations (2), magnetic induction at P due to this element is found to be ;

$$dB = \frac{\mu_0 N I a^2 dz}{2L[a^2 + (z_0 - z)^2]^{3/2}} \cdot$$

$\therefore$ The no. of turns in the element dz is NDZ/l. We note that $a/z_{0-z} = \tan\theta$ and $dz = ad\theta / \sin^2\theta$. Also $\sqrt{a^2 + (z_0 - z)^2} = a/\sin\theta$. Substituting these values, we get

$$dB = \frac{\mu_0 ni}{2L} \sin\theta d\theta$$

Summing up for all the elements, the total magnetic induction at P is found to be :-

$$B = \frac{\mu_0 ni}{2L} \int_{\theta_1}^{T-\theta_2} \sin\theta d\theta = \frac{\mu_0 ni}{2L}(\cos\theta_1 + \cos\theta_2)$$

At the midpoint of the solenoid, $\theta_1 = \theta_2$ and the magnetic induction is ;

$$B_{midpoint} = \frac{\mu_0 NI}{L} \cos\theta_1$$

If the selenoid is very long compared to its radius; $\theta_1 \rightarrow 0$, then:

$$B = \frac{\mu_0 NI}{l} = \mu_{0n} I; \quad \text{where} \quad n = \frac{N}{L}$$

which agrees with the result obtained on the basis of Amperes circuital law. Clearly the field at the centre of a finite solenoid is less than that for an infinite solenoid. If the ratio $L/a=14$, then the midpoint field is 99% of the field for an infinite solenoid.

If P is at one extreme end of the solenoid, $\cos\theta_2 = 0$ and the field at P becomes;

$$B_{end} = \frac{\mu_0 NI}{2L} \cos\theta_1$$

which is half the value of $B_{midpoint}$, when the solenoid is very long. Thus for a long solenoid, the magnetic field at one end is half that at the centre.

Magnetic Scalor Potential

We know that $\vec{\nabla} \times \vec{B} = \mu_0 \vec{j}$; for steady current.

If $\vec{j} = 0$ in some region of space, than in that region, $\vec{\nabla} \times \vec{B} = 0$.

$\therefore \vec{B}$ can be written as the gradient of a scalor potential.

$\therefore \vec{B} = -\mu_0 \vec{\nabla} \phi_m$

where $\phi_m \rightarrow$ magnetic scalor potential.

As $\vec{\nabla} \times \vec{B} = 0$, we have;

$$\Rightarrow D^2 \phi_m = 0$$

Thus ϕ_m satisfies Laplace equation.

Current loop in an external magnetic field

Consider a rectangular current loop abcd carrying a current I as shown.

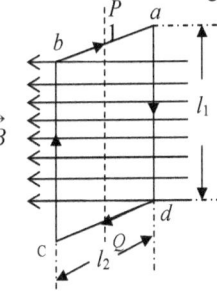

It is placed in a uniform magnetic field $\vec{B}$. The force acting on the side ad (of length l_1) of the loop is:-

$\vec{F_1} = I(\vec{l_1} \times \vec{B}) = Il_1 B$; where $\vec{l_1}$ is $\perp^r$ to $\vec{B}$. The direction of F_1 is along the $\perp^r$ to the plane containing $\vec{l_1} \& \vec{B}$.

The force F_2 acting on the side be will be equal and opposite to F_1. Similarly, the forces acting or the

sides ab and dc are equal and opposite. $\therefore$ the net force acting on the loop is zero, there is no translational motion of the loop. However, if the loop is free to rotate about the axis PQ, it will do so due to the torque produced by the forces acting on ad and bc. This torque will have the magnitude;

$\tau = F_1 l_2 \sin\theta$, where $\theta \to$ angle between $\vec{B}$ s the normal to the plane of the loop. $\therefore \tau = I l_1 l_2 B \sin\theta$

or, $\tau = ISB \sin\theta$; where $s = l_1 l_2 \to$ area of the loop.

or, $\vec{\tau} = I(\vec{S} \times \vec{B})$

If, $I\vec{S} = \vec{m} \to$ displacement of the loop; we have;

$\vec{\tau} = \vec{m} \times \vec{B}$

The torque τ tends to decrease the angle θ. If U is the potential energy of the loop, then;

$$\frac{\partial v}{\partial\theta} = \tau = mB\sin\theta,$$

a, $dU = mB\sin\theta d\theta \Rightarrow U = \int dU - mb \int \sin\theta d\theta = -mB\cos\theta + \text{const.}$

Assuming $\theta = \pi/2$ resembles U=0, we get const.=0.

$\therefore U = -mB\cos\theta = -\vec{m}.\vec{B}$

This expression is similar to the expression for potential energy of an electrostatic dipole of moment $\vec{P}$ placed in an electrostatic field $\vec{E}$:-

The above relationships are true for any loop of arbitrary shape.

Formal Theory of Magnetism :

Let $\vec{m_i} \to$ magnetic dipole moment of the ith atom.

$\vec{M} \to$ Magnetisation. The magnetization at any point is defined as the magnetic dipole moment per unit volume at that point.

$$\vec{M} \to \lim_{\Delta u \to 0} \sum_i \frac{\vec{m_i}}{\Delta u}$$

Also, $\vec{\nabla} \times \vec{M} \to \vec{j_m} \to$ magnetization current density (A/m^2).

Magnetic Intensity

In general, if the material is electrically conducting and in magnetized, then

$\vec{\nabla} \times \vec{B} = \mu_0 (\vec{j} + \vec{j_m})$; where $\vec{j} \to$ conduction current density involving transport of free carriers.

$\vec{j}_m \rightarrow$ circulatory current involving no charge transport.

$$\therefore \vec{\nabla} \times \vec{B} = \mu_0 \, (\vec{j} + \vec{\nabla} \times \vec{M})$$

$$\therefore \vec{\nabla} \times (\vec{B} - \mu_0 \, \vec{M}) = \mu_0 \, \vec{j}$$

or, $\vec{\nabla} \times \vec{H} = \vec{j}$; where $\vec{H} = \dfrac{\vec{B}}{\mu_0} - \vec{M}$

The above experience is known as Amperes circuited law inside matter.

$\vec{H} \rightarrow$ magnetic field intensity (SI unit : Amp/m).

Applying Stokes' theorem on $\vec{\nabla} \times \vec{H} = \vec{j}$, we get;

$$\iint_S (\vec{\nabla} \times \vec{H}) . \lambda ds = \oint_l \vec{H}.\vec{dl}$$

or, $\oint_l \vec{H}.\vec{dl} = \iint_S \vec{j}.\hat{\pi} \Rightarrow \oint_l \vec{H}.\vec{dl} = I$

$\qquad\qquad \rightarrow$ Integral form of Amperes circuital law inside matter.

So, we got; $\vec{H} = \dfrac{\vec{B}}{\mu_0} - \vec{M}$

$$\therefore \vec{B} = \mu_0 (\vec{H} + \vec{M})$$

In free space, $\therefore \vec{M} = 0 \Rightarrow \vec{B} = \mu_0 \, \vec{H}$, but in a magnetized material, $\vec{M} \neq 0$

$$\therefore \vec{B} = \mu_0 (\vec{H} + \vec{M})$$

The constitution relation :

The functional relation between $\therefore \vec{M} \, \& \, \vec{H}$ is known as magnetic constitutive relation.

Materials in which $\vec{M} \, \& \, \vec{H}$ share a linear relationship, are known as linear magnetic material. For isotropic linear materials, we have

$$\therefore \vec{M} = \chi_m \, \vec{H} ;$$ where $\chi_m \rightarrow$ dimensionless scalor quantity, known as magnetic susceptibility.

Linear isotropic materials $\begin{bmatrix} \chi_m > 0; \\ \chi_m < 0; \end{bmatrix}$

For $\chi_m > 0 \rightarrow$ paramagnetic materials; $\therefore \vec{H} \& \vec{M}$ are at the same direction

$\Rightarrow$ cause strengthening of $\vec{B}$ when placed in an external magnetic field.

For $\chi_m > 0 \rightarrow$ diamagnetic materials; $\therefore \vec{H} \& \vec{M}$ are at the opposite direction

$\Rightarrow$ cause weakening of $\vec{B}$ when placed in an external magnetic field.

Putting $\vec{M} = \chi_m \vec{H}$ in $\vec{B} = \mu_0(\vec{H} + \vec{M})$, we get:

$$\vec{B} = \mu_0 \vec{H} + \mu_0 \chi_m \vec{H})$$

or, $\vec{B} = \mu_0(1 + \chi_m)\vec{H} = \mu\vec{H}$

where $\mu = \mu_0(1 + \chi_m) \rightarrow$ magnetic permeability of the medium.

$\therefore \dfrac{\mu}{\mu_0} = 1 + \chi_m = \mu_r = Mr \rightarrow$ relative permeability.

For paramagnetic and diamagnetic materials, $|\chi_m| \ll 1; \Rightarrow \mu_r \approx 1$.

There is a class of magnetic materials which responds to external magnetic fields in a non-linear manner. That is, μ is a function of $\vec{H}$ and the linear relations; $\vec{M} = \chi_m \vec{H}$ and $\vec{B} = \mu\vec{H}$ do not hold. Such materials are known as <u>Ferromagnetic</u> materials. <u>Eg</u> iron. These materials possesses much higher degree of magnetization compared to paramagnetic materials. <u>Eg.</u> $\mu_r \approx 1$ for paramagnetic materials, but $\mu_r \approx 10^5$ for any ferromagnetic material. Ferromagnetic materials also exhibit on irreversible phenomena known as hysteresis, which makes such materials useful in constructing permanent magnets as transformers.

If $\vec{K}_m \rightarrow$ surface current density (bounded) in A/Mi $\therefore$ We have:-

$\vec{K}_m = \vec{M} \times \hat{en}$; where $\hat{en} \rightarrow$ unit vector normal to the surface.

Polarization density of the medium.

We have; $\vec{j} = \vec{\nabla} \times \vec{H}$ & $\vec{j}_m = \vec{\nabla} \times \vec{M}$

and $\vec{j}_m = \chi_m \vec{j}$; where $\chi_m = \mu_r - 1$

(P1) Region $0 \leq Z \leq 2m$ is occupied by an infinite slab of permeable material of $\mu_r = 2.5$. If $\vec{B} = 10y\,\hat{i} - 5x\,\hat{j}\ mWb/m^2$ within the slab, determine :-

$$\vec{j}, \vec{j}_m, \vec{M} \& \vec{K}_m \text{ on } z = 0$$

Solutions

By definition; $\vec{j} = \vec{\nabla} \times \vec{H} = \vec{\nabla} \times \dfrac{\vec{B}}{\mu}$

But $\vec{\nabla} \times \vec{B} = \begin{vmatrix} \hat{i} & \hat{j} & \hat{k} \\ \dfrac{\partial}{\partial x} & \dfrac{\partial}{\partial y} & \dfrac{\partial}{\partial z} \\ Bx & By & 0 \end{vmatrix}$ in Cartesian co-ordinate system.

$$= \left(\frac{\partial B_y}{\partial x} - \frac{\partial B_x}{\partial x} \right) \hat{k}$$

$$\vec{j} = \vec{\nabla} \times \frac{\vec{B}}{\mu_0 \mu_r}$$

$$\vec{j} = \vec{\nabla} \times \frac{\vec{B}}{\mu_0 \mu_r}$$

$$= \frac{1}{4\pi \times 10^{-7}(2.5)} \left(\frac{\partial B_y}{\partial x} - \frac{\partial B_x}{\partial y} \right) \hat{k} = \frac{10^6}{\pi}(-5-10).10^{-3}\,\hat{k} = -4.775\,\hat{k} = KA/m^2$$

$$\vec{j}_m = \chi_m \vec{j} = (\mu_r - 1)\vec{j} = 1.5(-4.775\,\hat{k}).10^3 = -7.163\,\hat{k}\ KA/m^2.$$

$$\vec{M} = \chi_m \vec{H} \chi_m \frac{\vec{B}}{\mu_0 \mu_r} = \frac{1.5(10y\,\hat{i} - 5x\,\hat{j})}{4\pi \times 10^{-7}(2.5)}.10^{-3} = 4.775y\,\hat{i} - 2.387x\,\hat{j}\ KA/m.$$

$\vec{K}_m = \vec{M} \times \hat{e}_n$ $\therefore$ Z=0 is lower side of the slab occupying $0 \leq Z \leq 2$, we have

$$\hat{e}_n = -\hat{k} \quad \therefore \vec{K}_m = (4.775y\,\hat{i} - 2.387x\,\hat{j}) \times \hat{x}$$

$$== 2.387x\,\hat{i} + 4.775y\,\hat{j}\ KA/m.$$

Magnetostatic Boundary Conditions

We define magnetostatic boundary conditions as the conditions that $\vec{H}(\text{or } \vec{B})$ field must satisfy at the boundary between two different media. From the figure below, we can derive :-

$\vec{B}_{1n} = \vec{B}_{2n}$, i.e., the normal component of $\vec{B}$ is continuous at the boundary. But as $\vec{B} = \mu \vec{H}$, we have:-

$$\mu_1 \vec{H}_{1n} = \mu_2 \vec{H}_{2n}$$

i.e., the normal component of $\vec{H}$ is discontinuous at the boundary.

Also, we have :-

$$\vec{H}_{1t} = \vec{H}_{2t} \Rightarrow \frac{\vec{B}_{1t}}{\mu_1} = \frac{\vec{B}_{2t}}{\mu_2}$$

$\Rightarrow$ The tangential component of $\vec{H}$ field in continuous, while that of the $\vec{B}$ field is discontinuous at the boundary – The law of reflection for magnetic flux lines at a boundary with nθ surface current is :-

$$\frac{\tan \theta_1}{\tan \theta_2} = \frac{\mu_1}{\mu_2}$$

ELECTROMAGNETIC INDUCTION:-

Laws of electromagnetic induction (Faraday's Law)

Faraday observed experimentally that whenever the magnetic flux linked with an electric circuit changed, a current was induced in the circuit – a phenomenon known as electromagnetic induction. The results of Faraday led to the development of two useful laws :-

1)   Neumann's law : The induced electromotive force (emf) in a circuit is equal to the time rate of change of the magnetic flux linked with the circuit.

2)   Leuz's Law : The direction of induced emf or current is such that it will oppose the change of flux producing it.

If $\phi \rightarrow$ flux linked with a circuit at time 't' (SI unit Wb).

$$\therefore \phi = \iint_S \vec{B}.\vec{n}\, dS$$

But $\vec{B} = \vec{\nabla} \times \vec{A}$, where $\vec{A} \rightarrow$ magnetic vector potential (Wb/m).

$$\therefore \phi = \iint_S (\vec{\nabla} \times \vec{A}).\vec{n}\,dS = \phi_l\, \vec{A}.\vec{dl} \text{ (by Stokes' Law)}.$$

The combination of two laws of electromagnetic induction yields;

$$\varepsilon_{ind} = -\frac{d\phi}{dt}; \quad \varepsilon_{ind} \rightarrow \text{induced emf (volts)}$$

The –ve sign indicates that ε_{ind} is also known as book emf.

If R$\rightarrow$ resistance of the circuit and i $\rightarrow$ induced current in Amps;

$$i = \frac{\varepsilon_{ind}}{R} = -\frac{1}{R}\frac{d\phi}{dt}$$

If the <u>electric field</u> is $\vec{E}$, the induced emf around a curve 'l' is given from definition;

$$\varepsilon_{ind} = \phi_l\, \vec{E}.\vec{dl}$$

But we have seen that $\varepsilon_{ind} = -\frac{d\phi}{dt}$ and $\phi = \iint_S \vec{B}.\vec{n}dS$.

$$\therefore \phi_l\, \vec{E}.\vec{dl} = -\frac{d}{dt}\iint_S \vec{B}.\hat{n}\,dS.$$

If the circuit is fixed, the time derivative can be moned inside the integral of the RHS, whence it becomes a partial derivative, Applying Stokes' theorem or the LHS, we get;

$$\iint_S (\vec{\nabla} \times \vec{E}).\hat{n} = \int_S \frac{\partial \vec{B}}{\partial t}.\vec{n}\,dS$$

The above is true for any arbitrary fixed surface S; $\therefore$ we get:-

$$\vec{\nabla} \times \vec{E} = -\frac{\partial \vec{B}}{\partial t} \rightarrow \text{differential form of Faraday's law}.$$

DISCUSSION

The phenomenon of electromagnetic induction is observed for the cases;

(1) A wire loop is in motion in the presence of a fixed magnetic field.

(2) A wire loop is fixed and the magnetic field is changing near it (or a magnet is in motion near the wire loop).

In both the above cases; $\varepsilon_{ind} = -\dfrac{\partial \phi}{\partial t}$; where $\phi = \displaystyle\iint_S \vec{B}.\vec{n}\,dS$.

For case (1); if the loop moves with a vel $\vec{v}$ in a fixed magnetic field, we can rightly assume that the free electrons (charges) on it also moves with velocity $\vec{v}$, giving rise to a magnetic force, $\vec{F}_m = q(\vec{v} \times \vec{B})$. This magnetic force $\vec{F}_m$ produces the induced emf ε_{ind}, whenever there is a cut or change in the magnetic lines of force. This ε_{ind} is also known as motional emf.

For case (2); as the loop is stationary, nθ magnetic force $\vec{F}_m$ is produced. Hence ε_{ind} is produced by a new type of a force. The field, which is the source of this new force is known as electric field (not an electrostatic field). Electrostatic force cannot generate emfs, but an entirely new kind of electric field produces a electric force, which can generate emf (rather induced emf). Evidently, a changing magnetic field induces an electric field (transformer emf)

Electrodynamics now involve three fields; $\vec{E}$ fields produced by stationary charges $(\vec{\nabla}.\vec{E} = \rho/\varepsilon_0; \; \vec{\nabla} \times \vec{E} = 0)$; $\vec{B}$ fields produced by electric currents $(\vec{\nabla}.\vec{B} = 0; \vec{\nabla} \times \vec{B} = \mu_0 \vec{j})$; and $\vec{G}$ fields (say) produced by changing magnetic fields $(\vec{\nabla}.\vec{G} = 0; \vec{\nabla} \times \vec{G} = \dfrac{\partial \vec{B}}{\partial t})$. Because $\vec{E}$ and $\vec{G}$ exert forces in the same way $\vec{F} = q(\vec{E} + \vec{G})$, it is tidier to regard their sum as a single entity and call the whole thing the "electric field". Hence for the static case, we have:-

$$\vec{\nabla}.\vec{E} = \rho/\varepsilon_0$$

$$\vec{\nabla} \times \vec{E} = 0$$

$$\vec{\nabla} \times \vec{B} = 0$$

$$\vec{\nabla} \times \vec{B} = \mu_0 \vec{j}$$

where $\vec{E} \rightarrow$ electrostatic field (N/C or V/m).
and for time varying fields (electrodynamics):-

$$\vec{\nabla}.\vec{E} = \rho/\varepsilon_0 \rightarrow \text{Gauss' law}$$

$$\vec{\nabla} \times \vec{E} = -\frac{\partial \vec{B}}{\partial t} \rightarrow \text{Faradays' law}$$

$$\vec{\nabla} \times \vec{B} = 0 \rightarrow \text{Gauss law in magnetism.}$$

$$\vec{\nabla} \times \vec{B} = \mu_0 \vec{j} \rightarrow \text{Amperes circuit al law.}$$

Where $\vec{E} \rightarrow$ electric field (N/C or V/M).

(P2)   A conducting bar can side freely over two conducting rais as shown. Calculate the induced voltage in the bar:-

(a) If the bar is stationary at y= $8\text{cm} \times \vec{B} = 4\cos 10^6 t\,\hat{k}\, mwb/m^2$.

(b) If the bar slides at a vol. $\vec{u} = 20\,\vec{j}\,ms^{-1} \,\&\, \vec{B} = 4\hat{k}\,mwb/m^2$.

(c) If the bar slides at a vol. $u = 20\,\vec{j}\,ms^{-1} \,\&\, \vec{B} = 4\cos(10^6 t - y)\hat{k}\,mwb/m^2$.

Solution :

(a) In this case, we have transformer end :-

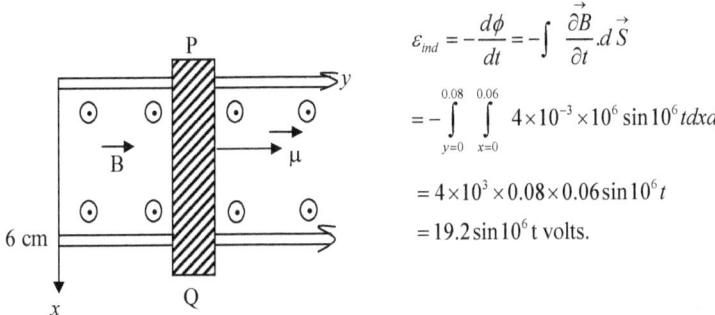

$$\varepsilon_{ind} = -\frac{d\phi}{dt} = -\int \frac{\partial \vec{B}}{\partial t} . d\vec{S}$$

$$= -\int_{y=0}^{0.08} \int_{x=0}^{0.06} 4 \times 10^{-3} \times 10^6 \sin 10^6 t \, dx dy$$

$$= 4 \times 10^3 \times 0.08 \times 0.06 \sin 10^6 t$$

$$= 19.2 \sin 10^6 t \text{ volts.}$$

This is the case of notional emf :-

$$\Delta\phi = \vec{B}.\vec{\Delta S} = B\,\Delta S; \text{ as the angle between } \vec{B} \,\&\, \vec{\Delta S} \text{ is } 0^\circ.$$

But $\Delta s = lu\,\Delta t$

$$\therefore \ \Delta\phi = B\,lu\Delta t \Rightarrow \frac{\Delta\phi}{\Delta t} = Blu$$

$$\therefore \ \varepsilon_{ind} = \frac{d\phi}{dt} = -\frac{\Delta\phi}{\Delta t} = Blu$$

$$\therefore \ \varepsilon_{ind} = -20(4 \times 10^{-3}) \times 0.06 = -4.8mV.$$

(c)   Both transformer emf and motional emf are present in this case.

$$\varepsilon_{ind} = -\frac{d\phi}{dt}; \quad \phi = \iint_S \vec{B}.d\vec{S}$$

$$\therefore \phi = \int_{x=0}^{0.06} \int_{y=0}^{y} 4\cos(10^6 t - y)dxdy = -4(0.06)\sin(10^6 t - y) \Big|_{y=0}^{y}$$

$$= -0.24\sin(10^6 t - y) + 0.24\sin 10^6 \, tmWb$$

But $y = \mu t = 20t$

$$\therefore \phi = -24\sin(10^6 t - 20t) + 0.24\sin 10^6 \, tMwb$$

$$\therefore \varepsilon_{ind} = -\frac{d\phi}{dt} \approx 240\cos(10^6 t - y) - 240\cos 10^6 t \text{ volts.}$$

Self Inductance

When a current flows in a circuit, the magnetic flux produced by the current depends on the geometry of the circuit and for nonferromagnetic materials, it is proportional to the current. Thus, if 'i' is the current flowing in a circuit, the magnetic flux can be written as;

$\phi = Li$, where L → self inductance or simply the inductance of the circuit.

(SI unit → Henry or H).

$$\therefore \frac{d\phi}{dt} = L\frac{di}{dt} + i\frac{dL}{dt}.$$

If a rigid stationary circuit is considered, then : $\dfrac{dL}{dt} = 0$

$$\therefore \frac{d\phi}{dt} = L\frac{di}{dt}.$$

But $\dfrac{d\phi}{dt} = \dfrac{d\phi}{di}.\dfrac{di}{dt} \Rightarrow L = \dfrac{d\phi}{di}$

Again, $\varepsilon_{ind} = -\dfrac{d\phi}{dt}$

$$\therefore \varepsilon_{ind} = -L\frac{di}{dt}$$

The –ve sign indicates that the induced emf opposes the flow of current. Thus the self inductance measures the ability of a circuit, to oppose the variation of current in it.

Calculation of Self-inductances

(1) Infinitely long solenoid :-

We have seen in pg-7 that B, for an infinitely long selenoid is :-

$$\vec{B} = \frac{\mu_0 N i}{l}$$

If A →area of cross section of the solenoid, the flux linking each turn is ;

$$Q_1 = BA = \frac{\mu_0 N i A}{l}$$

and the total flux linking N turns is :-

$$\phi = NQ_1 = \frac{\mu_0 N^2 A i}{l}$$

∴ self inductance of the solenoid is :-

$$L = \frac{\phi}{i} \Rightarrow L = \frac{\mu_0 N^2 A}{l}$$

(2)   Two infinitely long parallel wires:-

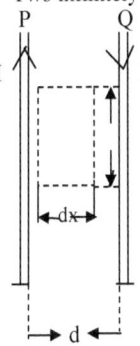

Let P and Q be two long parallel wires separated by a distinction 'd' in air. The lengths of the wires are much larger than d. The two wires carry the same current I in the opposite direction. The two wires thus constitute an open air T_x line. The radious of each wire 'a' is assumed to be much less than 'd', so that the flux inside the wires themselves can be neglected. The flux is concentrated in the region between the two wires. Consider a rectangular area of length 'l' and width 'dx' in the region between the wires. ∴ Flux ϕ through the rectangle is ;

$$d\phi = Bldx$$

If the rectangular element is at a distinction 'x' from the wire P, then:-

$$B = \frac{\mu_0 I}{2\pi}(\frac{1}{x} + \frac{1}{d-x})$$

where the contributions due to the wires P and Q are added. The total flux linkage is:-

$$\phi = \int\limits_{x=a}^{d-a} Bldx = \frac{\mu_0 Il}{2\pi}\left[l_n \frac{x}{x-d}\right]_a^{d-a} = \frac{\mu_0 Il}{\pi} l_n\left(\frac{d-a}{a}\right)$$

The self inductance, L is given by :-

$$L = \frac{\phi}{I} = \frac{d\phi}{dI} = -\frac{\mu_0 l}{\pi} l_n\left(\frac{d-a}{a}\right)$$

∴ $d \gg a$, the self inductance per unit length is :

$$\frac{L}{l} = \frac{\mu_0}{\pi} l_n \left(\frac{d}{a} \right)$$

If the wires are very close together, $(d-a) \approx a \Rightarrow L=0$

This is the principle of non-inductive winding employed.

(3)   Co-axial cylinders :-

Consider two coaxial cylinders of radii 'a' and 'b', respectively, with

(b) a as shown. The two cylinders carry the same current I in the opposite direction, thus forming a coaxial cable.

B at r>b.

If we assume an Amperian loop of radius r, the current enclosed by it is I-I=0. Thus, B=0 at r>b.

B at a< r < b:-

If we assume an Amperian loop of radius r, the current enclosed by it is I-I =0. Thus, B=0 at r > b.

B at a <r < b:-

Again, if we assume an Amperian loop of radius r, the current enclosed by it is I, due to the cylinder a. $\therefore$ by Amperes' law;

$$B = \frac{\mu_0 I}{2\pi r} .$$

If we imagine two co-axial cylinders of radii r and r+dr and of length 'l'

The flux in the region between the two cylinders is; $\vec{B}.l\vec{d}\,r = Bldr$

$$\therefore \phi = \int\limits_{r=a}^{b} Bldr = \frac{\mu_0 Il}{2\pi} l_n \int\limits_{a}^{b} dr/r = \frac{\mu_0 Il}{2\pi} \left(\frac{b}{a} \right) .$$

$\therefore$ the inductance L per unit length is :-

$$\frac{L}{l} = \frac{\phi}{l} l$$

$$\frac{L}{l} = \frac{\mu_0}{2\pi} l_n \left(\frac{b}{a} \right)$$

Mutual Inductance

Suppose we have N number of circuits, carrying currents. Let $\phi_{ij} \rightarrow$ flux through the ith circuit due to current in the jth circuit. For non-ferromagnetic materials, ϕ_{ij} is proportional to the current I_j of the jth circuit.

$$\therefore \phi_{ij} = M_{ij} I_j (i \neq j)$$

$M_{ij} \rightarrow$ mutual inductance between the circuits i and j (H). The mutual inductance between two circuits is defined as the flu9x linked with one circuit due to unit current in the other.

($\phi_{ii} \rightarrow$ self flux and $\dfrac{\phi_{ij}}{I_i} = L_i \rightarrow$ self inductance of the ith circuit).

The emf induced in the ith circuit due to the change of current in the jth circuit is :

$$\varepsilon_{ij} = -M_{ij} \frac{dI_{ij}}{dt}$$

The mutual inductance between two circuits may also be defined as the emf induced in one circuit. Due to the unit rate of change of current in the other. The to definitions lead to the same result when the fluxes are linearly related to the currents.

Calculation of Mutual Inductances

Two Solenoids : A short solenoid (s) is would over a long narrow solenoid (P) at its centre.

S → Secondary P → primary.

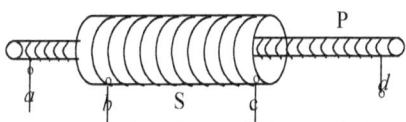
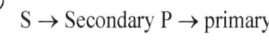

By connecting the terminals a and d to a source of electrical energy, a current is passed through the primary. If $N_1 \rightarrow$ lot no. of turns on the primary solenoid.

A → Cross sectional area and $l \rightarrow$ length, the magnetic flux linked with each turn of the secondary is :- ϕ= BA.

where $B = \dfrac{\mu N_1 I}{l}$ (from page – 7).

$\mu \rightarrow$ permeability of the core material.

If $N_2 \rightarrow$ No. of turns on the secondary solenoid, the flux linked with the secondary is:-

$$\phi_S = N_2 \phi = \frac{\mu N_1 N_2 AI}{l}$$

$\therefore$ Mutual inductance (M) between the P and S is ;

$$M = \phi_s / I = \frac{\mu N_1 N_2 A}{l}$$

If the current through P increases at the rate of dI/dt, the induced emf in S across the terminals b and c is - $M(dI/dt)$. this induced emf causes a flow of current in the secondary circuit (s) when the terminals b and c are connected through an external circuit.

Inductances in Series and Parallel

(1) Series Connection :-

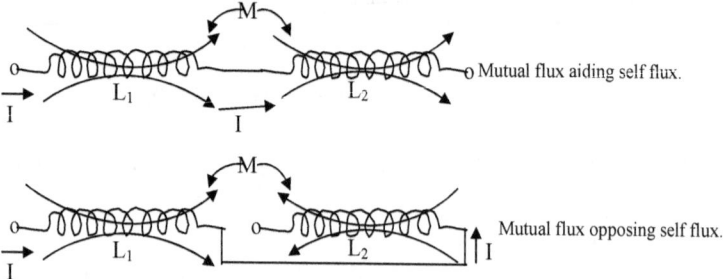

The above diagram is self explanatory.

Induced emf in coils 1 due to self inductance;

$$\varepsilon_{11} = -L_1 \frac{dI}{dt}$$

emf. Induced in coil 2 due to current I in coil 1 :-

$$\varepsilon_{21} = -M \frac{dI}{dt}$$

Induced emf in each 2 due to self inductance;

$$\varepsilon_{22} = -L_2 \frac{dI}{dt}$$

emf induced in coil 1 due to current I in coil 2 :-

$$\varepsilon_{12} = -M \frac{dI}{dt}$$

If all the fluxes aid one another, the lot emf is :-

$$\varepsilon = \varepsilon_{11} + \varepsilon_{22} + \varepsilon_{12} + \varepsilon_{21} = -(L_1 + L_2 + 2M) \frac{dI}{dt}$$

If Leq → equivalent inductance, then :-

$$\varepsilon = -L_{eq} \frac{dI}{dt}$$

$$\Rightarrow L_{eq} = L_1 + L_2 + 2M$$

If the mutual flux opposes the self flux, then :-

$$\varepsilon = \varepsilon_{11} + \varepsilon_{22} - \varepsilon_{12} - \varepsilon_{21}$$

$$\Rightarrow Leq = L_1 + L_2 - 2M$$

Thus in general $Leq = L_1 + L_2 \pm 2M$ in series.

If M=0, Leq = L$_1$ + L$_2$

(3)   Parallel connection

The aside fig is self explanatory.

$$\varepsilon_1 = -L_1 \frac{dI_1}{dt} - M \frac{dI_2}{dt}$$

$$\varepsilon_2 = -L_2 \frac{dI_2}{dt} - M \frac{dI_1}{dt}$$

$\therefore$ The two coils are in parallel, $\varepsilon_1 = \varepsilon_2 = \varepsilon$(say)

$$\therefore L_1 \frac{DI_1}{dt} + M \frac{DI_2}{dt} = -\varepsilon \text{ and } L_2 \frac{DI_2}{dt} + M \frac{DI_2}{dt} = -\varepsilon$$

Solving the above two equations, we get:-

$$\varepsilon = \frac{L_1 I_2 - M_2}{L_1 + L_2 - 2M} \frac{dI}{dt}.$$

But, $\varepsilon = -Leq \dfrac{dI}{dt}$ $\Rightarrow$ Leq=$\dfrac{L_1 L_2 - M_2}{L_1 + L_2 - 2M}$

If the mutual flux opposes the self flux, then :   Leq=$\dfrac{L_1 L_2 - M_2}{L_1 + L_2 + 2M}$

Thus, in general Leq=$\dfrac{L_1 L_2 - M_2}{L_1 + L_2 \pm 2M}$

If M=0, Leq=$L_1 L_2 \big/ L_1 + L_2$

Coefficient of Coupling

If we consider two loops, the mutual flux between them can be less than or at best equal to the self fluxes of the two loops.

$M = k\sqrt{L_1 L_2}$; where k $\rightarrow$ coefficient of coupling.

$0 \le k \le 1$; k $\rightarrow$ geometrical constant and it may be varied by varying the geometry, offering a means of constructing a variable inductance.

Magnetic Energy

The magnetic energy stored in a circuit is given by :-

$du = Id\phi$

for $\phi = LI \Rightarrow d\phi = LdI$.

$\therefore dU = LIDI$

when the current is increased, the magnetic energy stored in the circuit, is :-

$$U = \int dU = L\int_0^I IdI = \frac{1}{2}LI^2$$

also, $L = \dfrac{\phi}{I} \Rightarrow I = \dfrac{\phi}{l} \Rightarrow U = \dfrac{\phi^2}{2L}$.

Mathematical preliminaries
    (1)   Vector analysis.
    (2)   Co-ordinate systems.

Vector Analysis :

Vector : It is a physical quantity that has both magnitude and direction. If $\vec{A} \to$ vector, then in Cartesian co-ordinate systems; we have:-

$$\vec{A} = A_x\,\hat{i} + A_y\,\hat{j} + A_z\,\hat{k};\ \text{in terms of basis (or unit) vectors, } \hat{i}, \hat{j} \text{ and } \hat{k}.$$

$A_x, A_y, A_z \to$ components of $\vec{A}$.

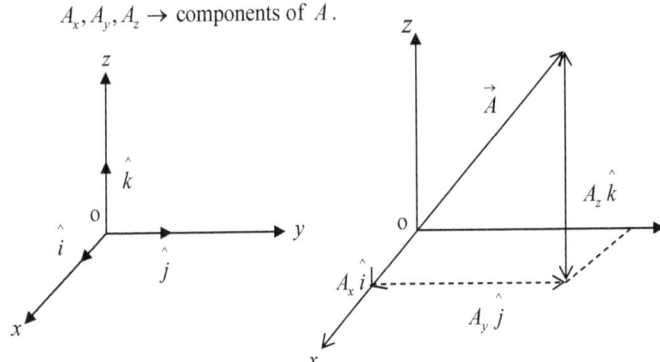

Unit or basis vectors :-

A vector $\vec{A}$ has both magnitude and direction. The magnitude $\vec{A}$ is a scalor quantity and written as $|\vec{A}|$. A unit vector $a\vec{A}$ along $\vec{A}$ is defined as a vector whose magnitude is unity and its direction is along $\vec{A}$, that is :-

$$a\vec{A} = \frac{\vec{A}}{|\vec{A}|};\ |a\vec{A}| = 1$$

Thus, $\vec{A} = |A|\,\hat{a}_n \to$ completely specifics $\vec{A}$ in terms of its magnitude $|\vec{A}|$ and its direction $a\vec{A}$.

    A vector $\vec{A}$ in Cartesian (or rectangular) co-ordinate system may be expressed as :

$$\vec{A} = A_x\,\hat{i} + A_y\,\hat{j} + A_z\,\hat{k};\ \text{where } A_x, A_y \text{ and } A_z \text{ are components of } \vec{A} \text{ along } x, y \text{ and } z. \text{ Directions}$$

respectively ; $\hat{i}, \hat{j}$ and $\hat{k}$ are unit vectors in the x, y and z directions respectively.

The magnitude of the vector $\vec{A}$ is given by;

$$|\vec{A}| = \sqrt{A_x^2 + A_y^2 + A_z^2}$$ and the unit vector along $\vec{A}$ is given by ;

$$\hat{a}_A = \frac{\vec{A}}{|A|} = \frac{A_x \hat{i} + A_x \hat{j} + A_x \hat{k}}{(A_x^2 + A_y^2 + A_z^2)^{1/2}}$$

Vector addition and subtraction:-

Two vectors $\vec{A}$ and $\vec{B}$ can be added together to give another vector $\vec{c}$;

$$\vec{c} = \vec{A} + \vec{B}.$$

The vector addition is carried out component by component. Thus;

If $\vec{A} = A_x \hat{i} + A_y \hat{j} + A_z \hat{k} \,\&\, \vec{B} = B_x \hat{i} + B_y \hat{j} + B_z \hat{k}$, then

$$\vec{c} = (A_x + B_x)\hat{i} + (A_y + B_y)\hat{j} + (A_z + B_z)\hat{k}$$

Vector subtraction is similarly carried out as ;

$$\vec{D} = \vec{A} - \vec{B} = \vec{A} + (-\vec{B})$$

$$= (A_x - B_x)\hat{i} + (A_y - B_y)\hat{j} + (A_z - B_z)\hat{k}$$

Graphically, vector addition and subtraction are obtained by either the parallelogram rule or head-to tail rule, which are performed exclusively at the high school levels.

All vectors obey the three basic laws of algebra;

Commutative law : $\vec{A} + \vec{B} = \vec{B} + \vec{A}$ (Addition); $K\vec{A} = \vec{A}K$(Multiplication)

Associative Law : $\vec{A} + (\vec{B} + \vec{c}) = (\vec{A} + \vec{B}) + \vec{c}$; $k(m\vec{A})$(Multiplication)

Distributive Law : $k(\vec{A} + \vec{B}) = K\vec{A} + K\vec{B}$

$$k, m \rightarrow \text{scalor constants.}$$

Position and distance vectors :-

A point P in Cartesian co-ordinates may be represented as $(x,\ y,\ z)$. The position vector $\vec{r}_p$ (or radius vector) of point P is the distance of P from the origin O and directed as OP. i.e.,

$$\vec{r}_p = \vec{OP} = x\hat{i} + Y\hat{j} + z\hat{k}.$$

The position vector of point P is useful in defining its position in space.

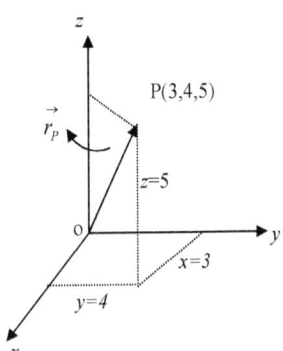

Point P(3, 4, 5), and its position vector is shown aside as an example. The position vector, $\vec{r}_p = 3\hat{i} = 4\hat{j} + 5\hat{k}$. If two points P and Q are given by $(x_1 y_1 z_1)$ and $(x_2 y_2 z_2)$ suspectively, then the distance vector (or separation vector) between P and Q is given by the displacement from P to Q, given as ;

$$\vec{r}_{pq} = \vec{r}_Q - \vec{r}_P$$

$$= (x_2 - x_1)\hat{i} + (y_2 - y_1)\hat{j} + (z_2 - z_1)\hat{k}.$$

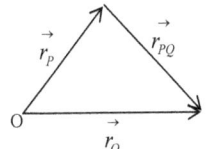

(PI) Given vectors $\vec{A} = \hat{i} - 3\hat{k}$ and $\vec{B} = 5\hat{i} - 2\hat{j} - 6$. Find :-

(a) $|\vec{A} + \vec{B}|$ (b) $5\vec{A} - \vec{B}|$ (c) The component of $\vec{A}$ along $\hat{j}$ (d) A unit vector parallel to (or along) $3\vec{A} + \vec{B}|$.

Solution (a) $|\vec{A} + \vec{B}| = 6\hat{i} + 2\hat{j} - 3\hat{k}$

$\therefore |\vec{A} + \vec{B}| = \sqrt{6^2 + 2^2 + (3)^2} = \sqrt{49} = 7$

(b) $5\vec{A} = 5\hat{i} + 15\hat{k}$

$\therefore 5\vec{A} - \vec{B} = (5\hat{i} + 15\hat{k}) - (5\hat{i} + 2\hat{j} - 6\hat{k}) = -2\hat{j} + 21\hat{k} = (0, -2, 21)$

(c) The component of $\vec{A}$ along $\hat{j}$ (y-axis) is $A_y = 0$

(d) Let $\vec{c} = 3\vec{A} + \vec{B} = (3\hat{i} + 9\hat{k}) + (5\hat{i} + 2\hat{j} - 6\hat{k})$

$$\hat{i} + 2\hat{j} + 3\hat{k}$$

A unit vector parallel to or along $\vec{c}$ is :-

$$\hat{a}_c = \frac{\vec{c}}{|c|} = \frac{\hat{i} + 2\hat{j} + 3\hat{k}}{\sqrt{1^2 + 2^2 + 3^2}} = \frac{\hat{i}}{\sqrt{14}} + \frac{2\hat{i}}{\sqrt{14}} + \frac{3\hat{k}}{\sqrt{14}} = 0.267\hat{i} + 0.535 + 0.802\hat{k} \text{ (Answer)}.$$

Note that $|\hat{a}_c| = 1$ as expected.

(P2) Points P and Q are located at (0, 2, 4) and (-3, 1, 5). Find :-

(a) The position vector $\vec{P}$. (b) The distance vector from $\vec{P}$ to $\vec{Q}$.

(b) The distance from (between) P and Q (d) A vector parallel to PQ with magnitude of 10.

Solution (a)   The position vector $\vec{P}$ of $\vec{r}_p = 2\hat{j} + 4\hat{k}$

  (b) $\vec{r}_{PQ} = \vec{r}_Q - \vec{r}_p = (3\hat{i} + \hat{j} + 5\hat{k}) - (2\hat{j} + 4\hat{k}) = -3\hat{i} - \hat{j} + \hat{k}$

  (c)   The distinction between P and Q is $d = |r_{PQ}|$.

$$d = \sqrt{(-3)^2 + (-1)^2 + 1^2} = \sqrt{11} = 3.317$$

  (d)   Set the registered vector is $\vec{A}$.

$\therefore \vec{A} = |A + \hat{a}_A$ where $|A| = 10$.

$\therefore \vec{A}$ is parallel to PQ, it must have the same unit sector as $\vec{r}_{PQ}$

$$\therefore \hat{a}_A = \frac{\vec{r}_{PQ}}{|r_{PQ}|} = \frac{-3i - j + k}{3.317} = -0.9045\,\hat{i} - 0.3015\,\hat{j} + 0.3015\,\hat{k}$$

$|a_A| = 1$ as usual.

$\therefore \vec{A} = |A| \hat{a}_A = 10\,\hat{a}_A = 9.045\,\hat{i} - 3.015\,\hat{j} + 3.015\,\hat{k}$.

Vector Multiplication :-

  (1) Scalor (or dot) product : $\vec{A}.\vec{B}$

  (2) Vector (or cross) product : $\vec{A} \times \vec{B}$

  (3) Scalor triple product : $\vec{A}(\vec{B} \times \vec{C})$

  (4) Vector triple product : $\vec{A} \times (\vec{B} \times \vec{C})$

Scalor (or dot) product :

$$\vec{A}.\vec{B} = |A||B|\cos\theta_{AB}$$

where $\theta_{AB} \to$ smaller angle between $\vec{A}$ and $\vec{B}$.

of $\vec{A} = A_x\hat{i} + A_y\hat{j} + A_z\hat{k}$ and $\vec{B} = B_x\hat{i} + B_y\hat{j} + B_z\hat{k}$

$\therefore \vec{A}.\vec{B} = A_xB_x + A_yB_y + A_zB_z \to$ a scalor quantity.

Two vectors $\vec{A}$ and $\vec{B}$ are said to be orthogonal or perpendicular to each other if $\vec{A}.\vec{B} = 0$

Laws :

1) Commutative law : $\vec{A}.\vec{B} = \vec{B}.\vec{A}$

2) Distributive law : $\vec{A}(\vec{B}+\vec{C}) = \vec{A}.\vec{B}+ \vec{A}.\vec{C}$

$$\vec{A}.\vec{A} = |\vec{A}|^2 = A^2$$

3) Also $\hat{i}.\hat{i} = \hat{j}.\hat{j} = \hat{k}.\hat{k} = 1$

and $\hat{i}.\hat{j} = \hat{j}.\hat{k} = \hat{k}.\hat{i} = 0$ or otherwise $=0$

(P3) prove the laws (1), (2) and (3).

Solution : Do it yourself

Vector (or cross) product :-

$$\vec{A}\times\vec{B} = |\vec{A}||\vec{B}|\sin\theta_{AB}\,\hat{a}_n$$

$\hat{a}_n \rightarrow$ unit vector normal to the plane containing $\vec{A}$ and $\vec{B}$.

If $\vec{A} = A_x i + A_y j + A_z k$ and $\vec{B} = B_x \hat{i}+ B_y \hat{j}+ B_z \hat{k}$

$$\therefore \vec{A}\times\vec{B} = \begin{vmatrix} \hat{i} & \hat{j} & \hat{k} \\ A_x & A_y & A_z \\ B_x & B_y & B_z \end{vmatrix} = (A_y B_z - A_z B_y)\hat{i} - (A_x B_z - A_z B_x)j + (A_x B_y - A_y B_x)\hat{k}.$$

which is obtained by "crossing" terms in cyclic permutation, hence the name cross product.

Basic laws:

1) It is not commutative : $\vec{A}\times\vec{B} \ne \vec{B}\times\vec{A}$

It is anti-commutative : $\vec{A}\times\vec{B} = -\vec{B}\times\vec{A}$

2) It is not associative : $\vec{A}\times(\vec{B}\times\vec{C}) \ne (\vec{A}\times\vec{B})\times\vec{C}$

3) It is distributive : $\vec{A}\times(\vec{B}\times\vec{C}) = (\vec{A}\times\vec{B})+(\vec{A}\times\vec{C})$

4) $\vec{A}\times\vec{A} = 0$; also:

$\vec{j}\times\hat{k} = \hat{i}$

$\vec{j}\times\hat{k} = \hat{i}$

$$\vec{k} \times \hat{i} = \hat{j}$$

(P4) Prove (1), (2), (3) and (4).

Solution : Do it yourself.

Throughout this note, we will use the right handed co-ordinate system, as we have drawn the first two figures.

Scalor triple product:-

$$\vec{A}.(\vec{B} \times \vec{C}) = \vec{B}.(\vec{C} \times \vec{A}) = \vec{C}.(\vec{A} \times \vec{B})$$

If $\vec{A} = A_x \hat{i} + A_y \hat{j} + A_z \hat{k}; \vec{B} = B_x \hat{i} + B_y \hat{j} + B_z \hat{k}; \vec{C} = C_x \hat{i} + C_y \hat{j} + C_z \hat{k}$

$$\therefore \vec{A}(\vec{B} \times \vec{C}) = \begin{vmatrix} A_x & A_y & A_z \\ B_x & B_y & B_z \\ C_x & C_y & C_z \end{vmatrix}$$

Vector Triple Product :-

$$\therefore \vec{A} \times (\vec{B} \times \vec{C}) = \vec{B} \times (\vec{A} \times \vec{C}) - \vec{C}(\vec{A} \times \vec{B})$$

$$\therefore (\vec{A}.\vec{B})\vec{C} \neq \vec{A}(\vec{B}.\vec{C}), \text{ but } (\vec{A}.\vec{B}).\vec{C} = \vec{C}(\vec{A}.\vec{B})$$

(P5) Find the angle between $\vec{A} = 3\hat{i} + 4\hat{j} + \hat{k}$ and $\vec{B} = 2\hat{j} - 5\hat{k}$

Solution : The angle θ_{AB} can be found by using ither dot product or cross product.

$$\vec{A}.\vec{B} = 8 - 5 == 3; |4| = \sqrt{3^2 + 4^2 + 1^1} = \sqrt{26} \qquad |\vec{B}| = \sqrt{2^2 + (-5)^2} = \sqrt{29}$$

$$\therefore \cos\theta_{AB} = \frac{\vec{A}.\vec{B}}{|A||B|} = \frac{3}{\sqrt{26.29}} = 0.1092$$

$$\therefore \cos\theta_{AB} = \cos^{-1}(0.1092) = 83.73°$$

Alternatively; $\vec{A} \times \vec{B} = \begin{vmatrix} \hat{i} & \hat{j} & \hat{k} \\ 3 & 4 & 1 \\ 0 & 2 & -5 \end{vmatrix} = -22\hat{i} + 15\hat{j} + 6\hat{k}$

$$\therefore |\vec{A} \times \vec{B}| = \frac{|\vec{A} \times \vec{B}|}{|A||B|} = \frac{\sqrt{745}}{\sqrt{26.29}} = 0.994$$

$$\therefore \theta_{AB} = \sin^{-1}(0.994) = 83.73°$$

(P6)   Let $\vec{E} = 3\hat{j} + 4\hat{k}$ $F = 4\hat{i} - 10\hat{j} + 5\hat{k}$

(a)   Find the component of $\vec{E}$ and $\vec{F}$ can be written as;

$$\vec{E}_F = |\vec{E}|\cos\theta_{EF}\,\hat{a}_F$$

$$|\vec{E}|\,|\hat{a}_F|\cos\theta_{EF}\,\hat{a}_F = (\vec{E}.\hat{a}_F)\hat{a}_F$$

But $\hat{a}_F = \dfrac{\vec{F}}{|\vec{F}|} = \dfrac{(\vec{E}.\vec{F})\vec{F}}{|\vec{F}|^2}$

Now, $\vec{E}.\vec{F} = 0 - 30 + 20 = -10$

$$|\vec{E}| = \sqrt{4^2 + (-10)^2 + 5^2} = \sqrt{141}$$

$$\therefore \vec{E}_F = -\frac{10}{141}(4\hat{i} - 10\hat{j} + 5\hat{k}) = -0.2837\hat{i} + 0.7092\hat{j} - 0.3546\hat{k}$$

(b)   Any vector which is $\perp^r$ to both $\vec{E}\,\&\,\vec{F}$ is given by ;

$$\vec{A} = \vec{E} \times \vec{F} = \begin{vmatrix} i & j & k \\ 0 & 3 & 4 \\ 4 & -10 & 5 \end{vmatrix} = 55\hat{i} + 16\hat{j} - 12\hat{k}$$

$\therefore$ Unit vector along $\vec{A}$ is ;

$$\vec{a}_A = \frac{\vec{A}}{|\vec{A}|} = \frac{55i + 16j - 12k}{\sqrt{55^2 + 16^2 + (-12)^2}} = 0.9398\hat{i} + 0.2734\hat{j} - 0.205\hat{k}$$

Note, $|\vec{a}_A| = 1$

(P6)   Prove that $\vec{A} = \dfrac{\vec{dA}}{dt} = A\dfrac{dA}{dt}$

Proof :

We know that $\vec{A}.\vec{A} = |A|^2$ $\therefore \dfrac{d}{dt}(\vec{A}.\vec{A}) = \dfrac{d}{dt}(A^2)$

or, $\vec{A}.\dfrac{d\vec{A}}{dt} + \vec{A}.\dfrac{d\vec{A}}{dt} = 2A\dfrac{dA}{dt} \Rightarrow 2\vec{A} = \dfrac{d\vec{A}}{dt} = 2A\dfrac{dA}{dt}$

$$\therefore \vec{A}.\frac{d\vec{A}}{dt} = A.\frac{dA}{dt} \text{ (proved)}$$

$\vec{A} \rightarrow$ constant;

If $\therefore \dfrac{d|\vec{A}|}{dt} = 0$

$\Rightarrow \vec{A}.\dfrac{d\vec{A}}{dt} = 0$

$\Rightarrow \vec{A}.\perp^r \dfrac{d\vec{A}}{dt}.$

VECTOR CALCULUS:-

Field $\rightarrow$ It is a function that specifies a particular quantity everywhere in a region.

Field can be a scalor or vector.

Equation of Scalor Field $\rightarrow$ Temperature distribution in a building, sound intensity in a theatre, electric potential in a region, etc.

Equation of Vector Field $\rightarrow$ Gravitational force on a body in space, electric force per unit charge in a region surrounding the stationary change.

GRADIENT :-

If $\phi \rightarrow$ a scalor field which can be differentiated, then from partial derivative theorem;

$$d\phi = \left(\dfrac{\partial \phi}{\partial x}\right)dx + \left(\dfrac{\partial \phi}{\partial y}\right)dy + \left(\dfrac{\partial \phi}{\partial z}\right)dz$$

or, $d\phi = \left(\hat{i}\dfrac{\partial \phi}{\partial x} + \hat{j}\dfrac{\partial \phi}{\partial y} + \hat{k}\dfrac{\partial \phi}{\partial z}\right).(\hat{i}\,dx + \hat{j}\,dy + \hat{k}\,dz$

Now, if $\vec{r} = x\,\hat{i} + y\,\hat{j} + z\,\hat{k} \rightarrow$ position vector.

$\therefore d\vec{r} = dx\,\hat{i} + dy\,\hat{j} + dz\,\hat{k}$

$\therefore d\phi = \vec{\nabla}\phi.d\vec{r}$

$\vec{\nabla}\phi = \hat{i}\dfrac{\partial \phi}{\partial x} + \hat{j}\dfrac{\partial \phi}{\partial y} + \hat{k}\dfrac{\partial \phi}{\partial z} \rightarrow$ gradient of ϕ, a vector quantity (in cortesian system).

$\vec{\nabla} \to$ del operatore $\equiv \hat{i}\dfrac{\partial}{\partial x} + \hat{j}\dfrac{\partial}{\partial y} + \hat{k}\dfrac{\partial}{\partial z}$ (in cartesian system)

Interpretation :

$$d\phi = \vec{\nabla}\phi.d\vec{r}$$

$$\Rightarrow d\phi = |\vec{\nabla}\phi\,||\,d\vec{r}\,|\cos\theta$$

$$\therefore d\phi_{max} = \vec{\nabla}\phi.d\vec{r}; \text{ for } \theta = 0°$$

Thus, 1) $\vec{\nabla}\phi$ points to the direction of maximum increase of ϕ.

2) $|\vec{\nabla}\phi|$ is the rate of increase of ϕ along the max increase.

(P7) If $\vec{r} = x\hat{i} + y\hat{j} + z\hat{k}$ find $\vec{\nabla}r$.

Solution : $r = |r| = \sqrt{x^2 + y^2 + z^2}$

$$\therefore \vec{\nabla}r = \hat{i}\frac{\partial r}{\partial x} + \hat{j}\frac{\partial r}{\partial x} + \hat{k}\frac{\partial r}{\partial z} = \hat{i}\frac{\partial}{\partial x}(x^2 + y^2 + z^2)^{1/2} + \hat{j}\frac{\partial}{\partial y}(x^2 + y^2 + z^2)^{1/2} + \hat{k}\frac{\partial}{\partial z}(x^2 + y^2 + z^2)^{1/2}$$

$$= \hat{i}\frac{1}{2}(x^2 + y^2 + z^2)^{-1/2}.2x + \hat{j}\frac{1}{2}(x^2 + y^2 + z^2)^{-1/2}.2y = \frac{x\hat{i} + y\hat{j} + z\hat{k}}{\sqrt{x^2 + y^2 + z^2}} = \frac{\vec{r}}{r} = \hat{r}$$

(P8) Given $\phi(x,y,z) = 3x^2 y - y^3 z^2, \vec{\nabla}\phi$ at $P(1,-2,1)$

Solution : $\vec{\nabla}\phi = \hat{i}\dfrac{\partial\phi}{\partial x} + \hat{j}\dfrac{\partial\phi}{\partial y} + \hat{k}\dfrac{\partial\phi}{\partial z}$

$$= 6xy\hat{i} + (3x^2 - 3y^2 z^2)\hat{j} - 2y^3 z\hat{k}$$

At $P(1,-2,-1); \vec{\nabla}\phi = -12\hat{i} - 9\hat{j} - 16\hat{k}$

DIVERGENCE:

The grade ϕ is :-

$$\vec{\nabla}\phi = \hat{i}\frac{\partial\phi}{\partial x} + \hat{j}\frac{\partial\phi}{\partial y} + \hat{k}\frac{\partial\phi}{\partial z}$$

or $\vec{\nabla}\phi = \left(\hat{i}\dfrac{\partial}{\partial x} + \hat{j}\dfrac{\partial}{\partial y} + \hat{k}\dfrac{\partial}{\partial z} \right)\phi$

$\vec{\nabla} = \hat{i}\dfrac{\partial}{\partial x} + \hat{j}\dfrac{\partial}{\partial y} + \hat{k}\dfrac{\partial}{\partial z} \rightarrow$ del operator in cartesian system.

If $\vec{A} = \hat{i}\,A_x + \hat{j}\,A_y + \hat{k}\,A_z$, then;

$\vec{\nabla}.\vec{A} = \dfrac{\partial}{\partial x}A_x + \dfrac{\partial}{\partial y}A_y + \dfrac{\partial}{\partial z}A_z \rightarrow$ divergence of $\vec{A}$, a scalor quantity (in Cartesian

co-ordinate system).

Interpretation:

$\vec{\nabla}.\vec{A}$ determines how much $\vec{A}$ spreads out (diverges) from the point.

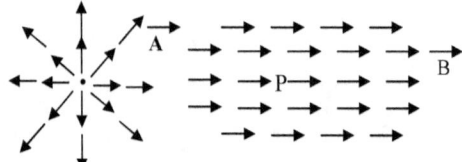

$\vec{A}$ at P has a large +ve divergence. If the arreas are pointed in towards P, then $\vec{\nabla}.\vec{A}$ is
−ve.

But $\vec{B}$ at P has $\vec{\nabla}.\vec{B} = 0$

$\therefore \vec{B}$ is selenoidal.

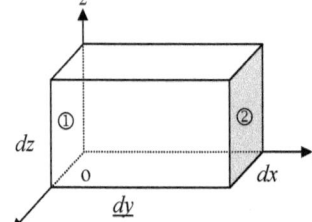

A fluid moves, so that its velocity at any point is
$v(x,y,z)$.

i.e., $\vec{v} = \hat{i}\,v_x + \hat{j}\,v_y + \hat{k}\,v_z$

We consider any part of the morning fluid as a cuboid in right handed Cartesian co-ordinate systems as shown in the fig.

Let $\vec{v}$ is at the origin 0.

From the fig., the volume element of the emboid is $dv=dxdydz$

$\dfrac{\partial v_x}{\partial x}, \dfrac{\partial v_y}{\partial y}, \dfrac{\partial v_z}{\partial z} \rightarrow$ represents the rates of change of the velocity components along the x, y & z axis respectively.

At the surface (1) : the velocity is v_y.

At the surface (5) : the velocity is $v_y + \dfrac{\partial v_y}{\partial x} dy$.

$\therefore$ The flux of vector $\vec{v}$ along the y-axis is :

$$(v_y + \dfrac{\partial v_y}{\partial y} dy)dxdz - v_y dxdz = \dfrac{\partial v_y}{\partial y} dxdydz$$

Similarly, the net outflow or flux vector $\vec{v}$ along the x-axis is :-

$\dfrac{\partial v_y}{\partial x} dxdydz$ and along the z-axis is $\dfrac{\partial v_z}{\partial z} dxdydz$.

$\therefore$ The total outflow or the flux of vector $\vec{v}$ from the volume element dv is :-

$$\left(\dfrac{\partial v_x}{\partial x} + \dfrac{\partial v_y}{\partial y} + \dfrac{\partial v_z}{\partial z} \right) dxdydz$$

$$= \left(\dfrac{\partial v_x}{\partial x} + \dfrac{\partial v_y}{\partial y} + \dfrac{\partial v_z}{\partial z} \right) dv$$

$$= \left(\vec{\nabla} . \vec{v} \right) dv$$

$\therefore$ The total outflow or per unit volume or the flux of $\vec{v}$ per unit volume from the elemental volume dv is :-

$$\vec{\nabla} . \vec{v}$$

If there is no gain of fluid from anywhere then :-

$\vec{\nabla} . \vec{v} = 0 \rightarrow$ equation of continuity for incompressible fluid.

In electrodynamics ($\Rightarrow$ current electricity);

$\vec{\nabla} . \vec{v} = 0 \rightarrow$ equation of continuity for incompressible fluid.

In electrodynamics ($\Rightarrow$ current electricity);

$\vec{v} \equiv \vec{j} \rightarrow$ current density $\Rightarrow$ current per unit area (A/m^2)

$\therefore$ For statical case; i.e., for DC;

$\vec{v} . \vec{j} = 0 \rightarrow$ equation of continuity in electrodynamics for DC.

(P9) a) If $\vec{r} = x\,\hat{i} + y\,\hat{j} + z\,\hat{k}$, find $\vec{\nabla}.\vec{r}$

  b) If $\vec{A} = \hat{k}$, find $\vec{\nabla}.\vec{A}$

Solutions :

$$\vec{r} = x\,\hat{i} + y\,\hat{j} + z\,\hat{k}$$

$$\vec{\nabla}.\vec{r} = \left(\hat{i}\frac{\partial}{\partial x} + \hat{j}\frac{\partial}{\partial y} + \hat{k}\frac{\partial}{\partial z}\right).\left(x\,\hat{i} + y\,\hat{j} + z\,\hat{k}\right)$$

$$\frac{\partial x}{\partial x} + \frac{\partial y}{\partial y} + \frac{\partial z}{\partial z} = 3$$

b) $$\vec{\nabla}.\vec{A} = \left(\hat{i}\frac{\partial}{\partial x} + \hat{j}\frac{\partial}{\partial y} + \hat{k}\frac{\partial}{\partial z}\right).\hat{k} = \frac{\partial}{\partial z} = 0$$

$\therefore$ $\vec{A}$ is solenoidal $\Rightarrow$ the flow of $\vec{A}$ is neigther created nor destroyed at any point $\rightarrow$ incompressible.

(P10) Determine 'a' such that $\vec{v} = (x+3y)\,\hat{i} + (y-2z)\,\hat{j} + (x+az)\,\hat{k}$ is solenoidal.

Solution : $$\vec{\nabla}.\vec{v} = \frac{\partial}{\partial x}(x+3y) + \frac{\partial}{\partial y}(y-2z) + \frac{\partial}{\partial z}(x+az)$$

  or, $\vec{\nabla}.\vec{v} = 1 + 1 + a = 2 + a$

  For $\vec{v}$ to be solenoidal, $\vec{\nabla}.\vec{v} = 0 \Rightarrow 2 + a = 0 \Rightarrow a = -2$

CURL :

  If $\vec{A} = \hat{i}\,Ax + \hat{j}\,Ay + \hat{k}\,Az$

  $\therefore$ Curl $\vec{A}$ is :

$$\vec{\nabla} \times \vec{A} = \left(\hat{i}\frac{\partial}{\partial x} + \hat{j}\frac{\partial}{\partial y} + \hat{k}\frac{\partial}{\partial z}\right) \times (\hat{i}\,Ax + \hat{j}\,Ay + \hat{k}\,Az)$$

  or, $$\vec{\nabla} \times \vec{A} = \begin{vmatrix} i & j & k \\ \dfrac{\partial}{\partial x} & \dfrac{\partial}{\partial y} & \dfrac{\partial}{\partial z} \\ Ax & Ay & Az \end{vmatrix}$$

Interpretation :

$\vec{\nabla} \times \vec{A}$ determines how much $\vec{A}$ rotates (curls) around the point.

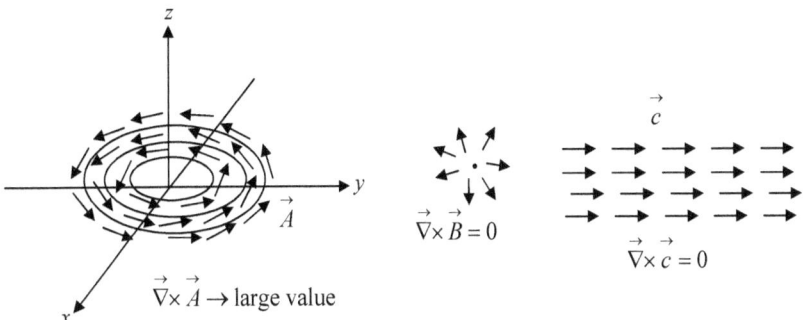

$\vec{\nabla} \times \vec{A} \rightarrow$ large value

$\vec{\nabla} \times \vec{B} = 0$

$\vec{\nabla} \times \vec{c} = 0$

$\vec{B}, \vec{C} \rightarrow$ irrotational vectors or potentials.

Important properties of ∇ (del) operator :-

$\vec{\nabla}(\vec{\nabla} \times \vec{A}); \vec{A} \rightarrow$ any vector

$\vec{\nabla} \times \vec{A} = \vec{c}$ (say). $\therefore \vec{c} \perp^{r}$ to both $\vec{A}$ and $\vec{\nabla}$.

$\therefore \vec{\nabla} . (\vec{\nabla} \times \vec{A}) = 0$

Conclusion : If there is any vector, which is solenoidal, i.e., its divergence is zero, then there exists another vector which has a curl relation with first vector.

Equation. If $\vec{B} \rightarrow$ solenoidal vector field, i.e., $\vec{\nabla} . \vec{B} = 0$

$$\Rightarrow \vec{B} = \vec{\nabla} \times \vec{A}$$

where $\vec{A} \rightarrow$ vector field (popularly known as potential function).

Practical example :

If $\vec{B} \rightarrow$ Magnetic Induction (SI unit Tesla, T), produced in a region due to some magnetic elements (e.g., current carrying wire), then, we have; $\vec{\nabla} . \vec{B} = 0$

hence, $\vec{B} = \vec{\nabla} \times \vec{A}$

where $\vec{A} \rightarrow$ magnetic vector potential.

2) $\vec{\nabla} \times (\vec{\nabla} \times \vec{A}); \vec{A} \to$ any vector

$$\vec{\nabla} \times \vec{A} = k \to \text{a scalor quantity.}$$

Now and $\vec{\nabla} \times \text{const} = 0$

$$\therefore \vec{\nabla} \times (\vec{\nabla} . \vec{A}) = 0$$

Conclusion : If there is any vector which is irrotational, i.e., its curl is zero, then there exists a scalor function (or field) which has a gradient relationship with the vector.

Equation. If $\vec{B} \to$ irrotational vector field, i.e., $\vec{\nabla} \times \vec{B} = 0$

$$\Rightarrow \vec{B} = \vec{\nabla} \phi$$

where $\phi \to$ scalor, popularly known as scalor potential.

Practical example :-

If $\vec{E} \to$ Electrostatic field (SI unit v/m) produced in a region due some electrostatic elements (e.g., point change at rest), then, we have :-

$$\vec{\nabla} \times \vec{E} = 0$$

hence $\vec{E} = -\vec{\nabla} \vee$

where $\vee \to$ electrostatic scalor potential (we will see the importance of $-$ve sign later).

Electrostatic and Gravitational fields are irrotational and their forces are conservative.

i.e., $\vec{\nabla} \times \vec{F}_G = 0 \ \& \ \vec{\nabla} \times \vec{F}_e = 0$ where $\vec{F}_G = \dfrac{Gm_1 m_2}{r^2} \hat{r}$

$\therefore \vec{F}_G = -\vec{\nabla} U_G \ \& \ \vec{F}_e = -\vec{\nabla}_{ve}$ & $\vec{F}_e = \dfrac{Kq_1 q_2}{r^2} \hat{r}$

$V_G \to$ Gravitational potential energy.

$V_e \to$ Electrostatic Potential energy.

Potential energy (and hence potential) exists only for conservative (irrotational) force fields.

Multiple $\vec{\nabla}$ (del) operators :-

$$\nabla^2 = \vec{\nabla}.\vec{\nabla} = \frac{\partial^2}{\partial x^2} + \frac{\partial^2}{\partial y^2} + \frac{\partial^2}{\partial z^2} \text{ in Cartesian – coordinate system.}$$

$\nabla^2 v = 0$ (Laplace equation)

$\nabla^2 v = -\rho/\varepsilon_o$ (Poisson equation)

$\nabla^2 v = \dfrac{1}{v^2}\dfrac{\partial^2 v}{\partial t^2}$ (wave equation)

$\nabla^2 v = \dfrac{1}{h^2}\dfrac{\partial v}{\partial t}$ (conduction of heat)

$\nabla^2 v = \dfrac{2m}{h v^2}(U-E)v$ (Schrodinger's wave equation)

(a)   If $\vec{A} \& \vec{B}$ are irrotational, prove that $\vec{A} \times \vec{B}$ is solenoidal.

(b)   If $\vec{A}$ is irrotational and $\vec{r} \rightarrow$, position vector, prove that $\vec{A} \times \vec{r}$ schenoidal.

Solution : (a) $\vec{A} \& \vec{B}$ are irrotational.

$\therefore \vec{\nabla} \times \vec{A} = 0 \,\& \, \vec{\nabla} \times \vec{B} = 0$

To find $\vec{A} \times \vec{B}$ is solenoidal or not, we have to compute $\vec{\nabla}(\vec{A} \times \vec{B})$

$\vec{\nabla}(\vec{A} \times \vec{B})$

$= \vec{B}(\vec{\nabla} \times \vec{A}) - \vec{A}.(\vec{\nabla} \times \vec{B})$ ....by vector identities.

$= 0$

$\Rightarrow \vec{\nabla}.(\vec{A} \times \vec{B}) = 0$

$\Rightarrow \vec{A} \times \vec{B} \rightarrow$ solenoidal (proved).

(b) $\vec{f} \rightarrow$ position vector, $\vec{r} = x\hat{i} + y\hat{j} + z\hat{k}$

As $\vec{A}$ is irrotational, $\vec{\nabla} \times \vec{A} = 0$

$\vec{\nabla}(\vec{A} \times \vec{r}) = \vec{r}(\vec{\nabla} \times \vec{A}) - \vec{A}(\vec{\nabla} \times \vec{r})$ ....by vector identities.

Now, $\vec{\nabla} \times \vec{r} = \begin{vmatrix} i & j & k \\ \dfrac{\partial}{\partial x} & \dfrac{\partial}{\partial y} & \dfrac{\partial}{\partial z} \\ x & y & z \end{vmatrix}$

$\therefore \vec{\nabla} \cdot (\vec{A} \times \vec{r}) = 0$

$\Rightarrow \vec{A} \times \vec{r} \equiv$ solenoidal (proved)

(P12) Let $\phi \rightarrow$ scalor function, which satisfies $\therefore \nabla^2 \phi = 0$. Prove that $\nabla^2 \phi$ is both solenoidal as well as irrotationa.

Solution : Do it yourself.

Integration of vectors :-

1) Line Integral
2) Surface Integral
3) Volume Integral

1) Line Integral :

Let $\therefore \vec{A} = Ax\,\hat{i} + Ay\,\hat{j} + Az\,\hat{k} \rightarrow$ any vector and $\vec{r} = x\,\hat{i} + y\,\hat{j} + z\,\hat{k} \rightarrow$ position vector in Cartesian co-ordinate system $d\vec{r} = dx\,\hat{i} + dy\,\hat{j} + dz\,\hat{k}$.

A curve 'l' is defined joining the points P_1 and P_2.

$\therefore \displaystyle\int_{P_1}^{P_2} \vec{A} \cdot d\vec{r} = \int_l \vec{A} \cdot d\vec{r}$ is defined as the line integral.

$\displaystyle\oint_l \vec{A} \cdot d\vec{r} = \oint_l (Axdx + Aydy + Azdz)$

If $\vec{A} = \vec{F}$ (force on a particle, moving along 'l'; the line integral represents the work done by the force.

If 'l' is a simple closed curve (i.e., it does nt intersect itself anywhere), the line integral becomes counter integral as ;

$$\oint_l \vec{A}.d\,\vec{r} = \phi(Axdx + Aydy + Azdz)$$

(P13) If $\vec{F}$ is irrotational (i.e., conservative), then prove that work done is moving a particle from the point $P_1(x_1y_1z_1)$ and $P_2(x_2y_2z_2)$ is independent of the path joining the two points.

Solution :

As $\vec{F}$ is irrotational or conservative, then $\vec{\nabla}\times\vec{F} = 0, \Rightarrow \vec{F} = \vec{\nabla}\phi$.

Where $\phi \rightarrow$ scalor function or field (potential).

Work done, $\vec{W} = \int_{P_1}^{P_2} \vec{F}.d\,\vec{r} = \int_{P_1}^{P_2} \vec{\nabla}\phi.d\,\vec{r}$ If $\vec{r} = x\hat{i} + y\,\hat{j} + z\hat{k}$

$$d\,\vec{r} = dx\hat{i} + dy\,\hat{j} + dz\hat{k}$$

$$\therefore W = \int_{P_1}^{P_2}\left(\hat{i}\frac{\partial\phi}{\partial x} + \hat{j}\frac{\partial\phi}{\partial y} + \hat{k}\frac{\partial\phi}{\partial z}\right).\left(\hat{i}\,dx + \hat{j}\,dy + \hat{k}\,dz\right)$$

or, $W = \int_{P_1}^{P_2}\frac{\partial\phi}{\partial x}dx + \frac{\partial\phi}{\partial y}dy + \frac{\partial\phi}{\partial z}dz = \int_{P_1}^{P_2} d\phi = \phi(P_2) - \phi(P_1)$

or, $W = \phi(x_2y_2z_2) - \phi(x_1y_1z_1)$

Clearly, W depends only on the points $P_1 = (x_1y_1z_1)$ and $P_2(x_2y_2z_2)$; and not on the path jointing them.

(P14) Given that $\vec{F} = (2xy + z^3)\hat{i} + x^2\,\hat{j} + 3xz^2\,\hat{k}$

    (a) Show that $\vec{F}$ is conservative (or irrotational).

    (b) Find the scalor potential.

    (c) Find the work done in moving an object in this force field from (1, -2, 1) to (3, 1, 4).

Solution : $\vec{F} = (2xy + z^3)\hat{i} + x^2\,\hat{j} + 3xz^2\,\hat{k}$

(a) Now, $\vec{\nabla}\times\vec{F} = \begin{vmatrix} i & j & k \\ \dfrac{\partial}{\partial x} & \dfrac{\partial}{\partial y} & \dfrac{\partial}{\partial z} \\ 2xy + z^3 & x^2 & 3xz^2 \end{vmatrix} = \dfrac{\partial}{\partial y}(3xz^2) - \dfrac{\partial}{\partial z}(x^3)\hat{i} -$

$$\frac{\partial}{\partial x}(3xz^2) - \frac{\partial}{\partial z}(2xy + z^3)\hat{j} +$$

$$\frac{\partial}{\partial x}(x^2) - \frac{\partial}{\partial y}(2xy + z^3)\hat{k}$$

$$= (3z^2 - 3z^2)\hat{j} + (2x - 2x)\hat{k}$$

$$= 0$$

$\therefore \vec{F}$ is conservative or irrotational.

(b) $\therefore \vec{\nabla} \times \vec{F} = 0 \Rightarrow \vec{F} = \vec{\nabla}\phi$; $\phi \rightarrow$ scalor potential.

$$\Rightarrow (2xy + z^3)\hat{i} + x^2\hat{j} + 3xz^2\hat{k} = \hat{i}\frac{\partial\phi}{\partial x} + \hat{j}\frac{\partial\phi}{\partial y} + \hat{k}\frac{\partial\phi}{\partial z}$$

$$\frac{\partial\phi}{\partial x} = 2xy + z^3; \frac{\partial\phi}{\partial y} = x^2; \frac{\partial\phi}{\partial z} = 3xz^2$$

$$\Rightarrow \phi = x^2 y + xz^3 + f; \ \phi = x^2 y + q; \ \phi = xz^3 + w$$

The above three satisfies simultaneously, if ;

$f = 0, g = xz^3, h = x^2 y.$

Thus, with any one of the above;

$\phi = x^2 y + xz^3.$

(c)   Work done ;

$$W = \phi(P_2) - \phi(P_1)$$

or, $W = \phi(3,1,4) - \phi(1,-2,1)$

or, $W = (3^2.1 + 3.4^3) - (1^2.(-2) + 1.1^2)$

Surface Integral :

Let 'S' be a two sided surface. If a differential surface areas is associated by a vector $\vec{ds}$, such that $\vec{ds} = \hat{n}\vec{ds}$, then,

$\iint\limits_{S} \hat{n}ds$ represents surface integration.

$(\vec{A} = Ax\hat{i} + Ay\hat{j} + Az\hat{k})$

$\iint\limits_{S} \vec{A}\hat{n}ds$ is the flux of $\vec{A}$ overs.

$\iint\limits_{S} \vec{A}\hat{n}ds$ represents integration of $\vec{A}$ over a closed surface.

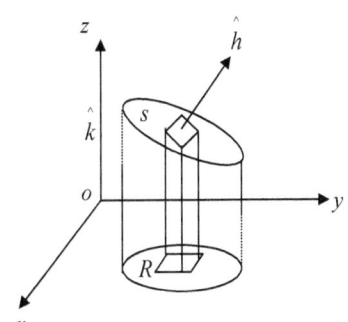

For an arbitrary surface 's' as shown, with projection 'R' on the xy plane, we have ;

$$\iint_S \vec{A}\,\hat{n}\,ds = \iint_R \vec{A}.\hat{n}\,\frac{dxdy}{|\hat{n}.\hat{k}|}$$

(P15) If If $\vec{A} = 4xz\,\hat{i} - y^2\,\hat{j} + yz\,\hat{k}$, evaluate $\iint_R \vec{A}.\hat{n}\,dS$ over the surface of a cube bounded by $x=0$ to $x=1$; $y=0$ to $y=1$; $z=0$ to $z=1$.

Solution :

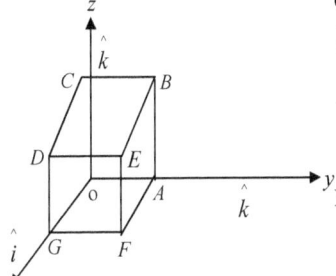

On the surface DEFG;

$\hat{n} = \hat{i}, x = 1$

$$\therefore \iint_{DEFG} \vec{A}.\hat{n}\,ds = \int_0^1\int_0^1 (4z\,\hat{i} - y^2\,\hat{j} + yz\,\hat{k}).\hat{i}\,dydz$$

$$\text{DEFG} = \int_0^1\int_0^1 4z\,dydz = 2$$

On ABCO; $\hat{n} = \hat{i}$ and $x = 0$

$$\therefore \iint_{ABCO} \vec{A}.\hat{n}\,ds = \int_0^1\int_0^1 (-y^2\,\hat{j} + yz\,\hat{k}).(-\hat{i})\,dydz = 0$$

On ABEF; $\hat{n} = \hat{j}$, y=0

$$\therefore \iint_{ABEF} \vec{A}.\hat{n}\,ds = -1$$

On OGDC; $\hat{n} = -\hat{j}$, y=0

$$\therefore \iint_{AGDC} \vec{A}.\hat{n}\,ds = 0$$

On BCDE; $n = \hat{k}$, $z=1$

$$\therefore \iint\limits_{BCDE} \vec{A}.\hat{n}\,ds = \frac{1}{2}$$

On AGFO; $\hat{n} = -k, z = 0$

$$\therefore \iint\limits_{AGFO} \vec{A}.\hat{n}\,ds = 0$$

$$\therefore \text{Adding } \iint\limits_{S} \vec{A}.\hat{n}\,ds = 2+0+(-1)+0+\frac{1}{2}+0 = \frac{3}{2}.$$

Volume Integrals :-

If a closed surface in space encloses a volume V, then;

$$\therefore \iiint\limits_{V} \vec{A}\,dV \rightarrow \text{volume integral (also known as space integral)}.$$

(P16) If $\vec{A}=2xz\,\hat{i}-x\,\hat{j}+y^2\,\hat{k}$ compute $\iint\limits_{V} \hat{A}\,dV$ in the space (or volume) bounded by the surfaces $x = 0, x = 2; y = 0, y = 6; z = x^2, z = 4$

Solution :

$$\therefore \iiint\limits_{V} A\,dV = \int\limits_{0}^{2}\int\limits_{0}^{6}\int\limits_{x^2}^{4}(2xz\,\hat{i}-x\,\hat{j}+y^2\,\hat{k})dxdydz$$

$$= \hat{i}\int\limits_{0}^{2}\int\limits_{0}^{6}\int\limits_{x^2}^{4}(2xzdxdydz - \hat{j}\int\limits_{0}^{2}\int\limits_{0}^{6}\int\limits_{x^2}^{4}xdxdydz + \hat{k}\int\limits_{0}^{2}\int\limits_{0}^{6}\int\limits_{x^2}^{4}y^2dxdydz\,128\,\hat{i}-24\,\hat{j}+384\,\hat{k}$$

The Integral Theorems :

1) The divergence theorem (by Gauss).

If a Vol. V is bounded by closed surface S, & $\vec{A}$ is a vector function of position with continuous derivative, then :-

$$\iint\limits_{S} \vec{A}.\hat{n}\,ds = \iiint\limits_{V} (\vec{\nabla}.\vec{A})dV.$$

Interpretation :

The RHS determines how much $\vec{A}$ diverges through a total vol.V. As per the divergence theorem, this must be equal to flux of $\vec{A}$ passing through the closed surface S which encloses the vol. V. Quantitatively, it can be stateted that;

(Flow of $\vec{A}$ per unit surface area) × (total surface area)

= (spreading of $\vec{A}$ per unit vol. Enclosed by the surface) × (total volume).

2)   The Stokes' theorem :

If S is an open two sided surface bounded by a simple closed curve '*l*' (non-intersecting) and if $\vec{A}$ has continuous derivative, then :-

$$\int_l \vec{A}.\vec{dl} = \iint_S (\vec{\nabla}.\vec{A})\hat{n}\,dS.$$

Interpretation :-

$$\iint_S (\vec{\nabla}.\vec{A})\hat{n}\,dS.$$

$$\int_l \vec{A}.\vec{dl}$$

$\vec{\nabla} \times \vec{A}$ measures the twist of $\vec{A}$. Now, the integral of $\vec{\nabla} \times \vec{A}$ over some surface or the flux of $\vec{\nabla} \times \vec{A}$ through that surface represents the total amount of "swirl". This "swirl" can be determined by all around the edge of the surface bounded by a simple curve. This is state by Stoke's theorem.

7)   Repeat problem (P15) using the divergence theorem :

Solution : Tonevaluate $\iint_S \vec{A}.\hat{n}\,ds = \iiint_V (\vec{\nabla}.\vec{A})dV$

$$(\vec{\nabla}.\vec{A}) = \frac{\partial}{\partial x}(4xz) - \frac{\partial}{\partial y}(y^2) + \frac{\partial}{\partial z}(yz) = 4z - y.$$

$$\therefore \iiint\limits_{V} (\vec{\nabla}.\vec{A})dV = \int\limits_{x=0}^{1} \int\limits_{y=0}^{1} \int\limits_{z=0}^{1} (4z-y)dxdydz.$$

$$\int\limits_{x=0}^{1} \int\limits_{y=0}^{1} 2z^2 - yz \mid_{z=0}^{1} dxdy = \int\limits_{x=0}^{1} \int\limits_{y=0}^{1} (2-y)dxdy = \int\limits_{x=0}^{1} 2y - \frac{1}{2}y^2 \mid_{y=0}^{1} dx$$

$$= \frac{3}{2} \int\limits_{x=0}^{1} dx = \frac{3}{2} \ \ (\text{Answer})$$

(P18) A continuous charge distribution of volume charge density $P(x,y,z,t)$ moves with a velocity $\vec{V}(x,y,z,t)$. If there are no sources or sinks, prove that, $\vec{\nabla}.\vec{j} + \dfrac{\partial \rho}{\partial t} = 0$; where $\vec{j} = \rho \vec{v}$, which is current density (A/m^2).

Solution :

$\vec{j} \rightarrow$ current density is a vector quantity (A/m^2).

Its magnitude is equal to the amount of charge passing per unit area per unit time, through a surface element at right angles to the flow. Its direction is along the motion of the charge.

$\therefore$ The current due to the flow of charge is :-

$$I = \iint\limits_{S} \vec{j}.\hat{n}\, ds$$

By divergence theorem; $\oiint\limits_{S} \vec{j}.\hat{n}\, ds = \iiint\limits_{V} (\vec{\nabla}.\vec{j})dV$

But, as the charge is conserved and also there is no sources or sin   whatever flows out through the surface must be due to rate of change of charge density inside.

$$\therefore \iiint\limits_{V} (\vec{\nabla}.\vec{j})dV = -\frac{d}{dt} \iiint\limits_{V} \rho dV = -\iiint\limits_{V} \frac{\partial \rho}{\partial t}dV$$

-ve sign implies that an outward flow decreases the charge left in the volume dV.

$\therefore$ The above applies for any volume, we get :-

$$\vec{\nabla}.\vec{j} + \frac{\partial \rho}{\partial t} = 0 \rightarrow \text{equation of eontinuity.}$$

For statical case, $\dfrac{\partial \rho}{\partial t} = 0$

$$\therefore \vec{\nabla} . \vec{j} = 0$$

$\Rightarrow \vec{j}$ is solenoidal. The flow of $\vec{j}$ is neither created nor destroyed at any point within our consideration.

Let $\vec{F} \rightarrow$ a conservative (or irrotational) force.

$$\therefore \vec{\nabla} \times \vec{F} = 0$$

By Stokes' theorem;

$$\oint_l \vec{F} . \vec{dl} = \iint_S (\vec{\nabla} \times \vec{F}) . \hat{n} \, ds$$

$$\therefore \oint_l \vec{F} . \vec{dl} = 0$$

i.e., work done by a conservative force in a closed loop (path) is zero. We conclude for a conservative force $\vec{F}$;

(1) The work done by $\vec{F}$ in moving a charge from one point to another is independent of the path through which the charge (or mass) is traversed and it depends only on the initial and final co-ordinates of the charge (or mass).

(2) The work done is a closed loop (path) on a particle (charge or mass) by $\vec{F}$ is zero.

(3) There exists a scalor potential (or potential energy) equivalent to work done $\vec{F}$, such that $\vec{F} = -\vec{\nabla} U$.

Co-ordinate Systems :

In EM theory and Antenna, we need to solve problems in co-ordinate systems other than cartesian.

A point or a vector can be represented in any curvilinear co-ordinate system, which may be orthogonal or non-orthogonal.

An orthogonal system is one in which the co-ordinats are mutually perpendicular.

Non-orthogonal systems are difficult to work with and they are of little or no practical use. Examples of orthogonal co-ordinate systems include Cartesian, the circular cylindrical, the spherical, the elliptic cylindrical,the parabolic cylindrical, the council, the prolate spheroidal, the oblate spheroidal and the ellipsoidal. A considerable amount of work and time may be saved by choosing a co-ordinate system that both fits the given problem. A difficult problem is one co-ordinate system may turn out to be easy in another system.

In this section, we discuss about three orthogonal co-ordinate systems :-

1) Rectangular Cartesian co-ordinate system (x, y, z).

2) Circular cylindrical co-ordinate system (r, ϕ, z)

3) Spherical co-ordinate systems (r, θ, ϕ)

Cartesian (rectangular) co-ordiante system :-

$\hat{i}, \hat{j}, \hat{k} \rightarrow$ unit vectors (basis vectors) along the x, y & z –axis respectively.

Any vector, $\vec{A} = \hat{i} \, Ax + \hat{j} \, Ay + \hat{k} \, Az$

$$\hat{i}.\hat{i} = \hat{j}.\hat{j} = \hat{k}.\hat{k} = 1$$

$$\hat{i}.\hat{i} = \hat{j}.\hat{j} = \hat{k}.\hat{k} = 0$$

Line element :-

$$\vec{dl} = \hat{i} \, dx + \hat{j} \, dy + \hat{k} \, dz$$

$$\therefore \therefore \hat{dl} = dx^2 + dy^2 + dz^2$$

Area element :-

$$ds = dxdy$$

Volume element :

$$dV = dxdydz$$

Gradient :

$$\vec{\nabla} V = \hat{i} \frac{\partial V}{\partial x} + \hat{j} \frac{\partial V}{\partial y} + \hat{k} \frac{\partial V}{\partial z} \qquad \vec{\nabla} = \hat{i} \frac{\partial}{\partial x} + \hat{j} \frac{\partial}{\partial y} + \hat{k} \frac{\partial}{\partial z}$$

Divergence :

$$\vec{\nabla}.\vec{E} = \frac{\partial}{\partial x}Ex + \frac{\partial}{\partial y}Ey + \frac{\partial}{\partial z}Ez$$

Curl;

$$\vec{\nabla} \times \vec{E} = \begin{vmatrix} i & j & k \\ \dfrac{\partial}{\partial x} & \dfrac{\partial}{\partial y} & \dfrac{\partial}{\partial z} \\ x & y & z \end{vmatrix}$$

Laplacian Operator :

$$\nabla^2 V = \frac{\partial^2 V}{\partial x^2} + \frac{\partial^2 V}{\partial y^2} + \frac{\partial^2 V}{\partial z^2}$$

Circular cylindrical co-ordinate system :-

P(r, ϕ, z) $\hat{e}_r, \hat{e}_\phi, \hat{e}_z \rightarrow$ unit vectors.

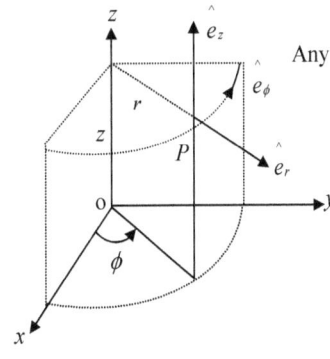

Any vector $\vec{A}$ can represented as ;

$$\vec{A} = \hat{e}_r A_r + \hat{e}_\phi A_\phi + \hat{e}_z.A_z.$$

$$\hat{e}_r.\hat{e}_r = \hat{e}_\phi.\hat{e}_\phi = \hat{e}_z.\hat{e}_z = 1$$

$$\hat{e}_r \times \hat{e}_r = \hat{e}_\phi \times \hat{e}_\phi = \hat{e}_z \times \hat{e}_z = 0$$

Live element :-

$$\vec{dl} = \hat{e}_r.dr + \hat{e}_\phi \, rd_\phi + \hat{e}_z \, dz$$

$$\Rightarrow dl^2 = dr^2 + r2\hat{e}_\phi \, rd_\phi + \hat{e}_z \, dz$$

Area element :

$$ds = rd\phi dz$$

Volume element :-

$$dv = rd\phi dzdr$$

$$\vec{\nabla} \equiv \hat{e}_r \frac{\partial}{\partial r} + \hat{e}_\phi \frac{1}{r}\frac{\partial}{\partial \phi} + \hat{e}_z \frac{\partial}{\partial z}$$

Gradient; $\vec{\nabla} V \equiv \hat{e}_r \dfrac{\partial V}{\partial r} + \hat{e}_\phi \dfrac{1}{r}\dfrac{\partial V}{\partial \phi} + \hat{e}_z \dfrac{\partial V}{\partial z}$.

Divergence :

$$\vec{\nabla}.\vec{E} \equiv \dfrac{1}{r}\dfrac{\partial}{\partial r}(rE_r) + \dfrac{\partial}{\partial \phi}E_\phi.\dfrac{1}{r} + \dfrac{\partial}{\partial z}E_z$$

Curl;

$$\vec{\nabla}.\vec{E} = \dfrac{1}{r}\begin{vmatrix} e_r & e_\phi & e_z \\ \dfrac{\partial}{\partial r} & \dfrac{\partial}{\partial \phi} & \dfrac{\partial}{\partial z} \\ E_r & rE_\phi & E_z \end{vmatrix}$$

Lapalacian Operator :

$$\nabla.V = \dfrac{1}{r}\dfrac{\partial}{\partial r}(r\dfrac{\partial V}{\partial r}) + \dfrac{1}{r^2}\dfrac{\partial^2 V}{\partial \phi^2}\dfrac{\partial^2 V}{\partial z^2}$$

The relationships between the variables (x, y, z) of the certesian co-ordinate system and that of the cylindrical system (r, ϕ, z) are :-

$(x, y, z) \rightarrow (r, \phi, z)$:-

$r = \sqrt{x^2 + y^2}$; $\phi = \tan^{-1}\dfrac{y}{x}$; $z = z$

        or

$(r, \phi, z) \rightarrow (x, y, z):-$

$x = r\cos\phi$; $y = r\sin\phi$; $z = z$

Unit vector transformation;

$\hat{e}_r = \cos\phi\,\hat{i} + \sin\phi\,\hat{j}$ $\hat{i} = \cos\phi\,\hat{e}_r - \sin\hat{e}_\phi$

$\hat{e}\phi = -\sin\phi\,\hat{i} + \cos\phi\,\hat{j}$ $\hat{j} = \sin\phi\,\hat{e}_r + \cos\phi\,\hat{e}_\phi$

$\hat{e}z = \hat{k}$ $\hat{k} = \hat{e}_z$

Spherical co-ordinate system :-

$$P(r,\theta,\phi) \qquad\qquad \hat{e}_r, \hat{e}_\theta, \hat{e}_\phi \rightarrow \text{unit vectors.}$$

Any vector x $\vec{A}$ can be represented as;

$$\vec{A} = \hat{e}_r A_r + \hat{e}_\theta A_\theta + \hat{e}_\phi A_\phi = 1$$

and, $\hat{e}_r.\hat{e}_r = \hat{e}_\theta.\hat{e}_\theta = \hat{e}_\phi.\hat{e}_\phi = 1$

& $\hat{e}_r \times \hat{e}_r = \hat{e}_\theta \times \hat{e}_\theta = \hat{e}_\phi \times \hat{e}_\phi = 0$

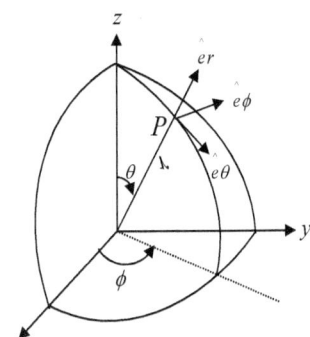

Since element :

$$\vec{dl} = \hat{e}_r\, dr + \hat{e}_\theta\, rd\theta + \hat{e}_\phi\, r\sin\theta\phi$$

$$\therefore dl^2 = dr^2 + r^2 d\theta^2 + r^2 \sin^2\theta d\phi^2$$

Area element; $ds = r^2 \sin\theta d\theta d\phi$.

Volume element; $dV = r^2 \sin\theta d\theta d\phi dr$

$$\vec{\nabla} \equiv \hat{e}_r \frac{\partial}{\partial r} + \hat{e}\theta \frac{1}{r}\frac{\partial}{\partial\theta} + \hat{e}\theta \frac{1}{r\sin\theta}\frac{\partial}{\partial\theta}$$

Gradient :

$$\vec{\nabla} V \equiv \hat{e}_r \frac{\partial V}{\partial r} + \hat{e}\theta \frac{1}{r}\frac{\partial V}{\partial\theta} + \hat{e}\theta \frac{1}{r\sin\theta} - \frac{\partial V}{\partial\theta}.$$

Divergence;

$$\vec{\nabla}.\vec{E} \equiv \frac{1}{r^2\sin\theta}\left[\sin\theta\frac{\partial}{\partial r}(r^2 E_r) + r\frac{\partial}{\partial\theta}(\sin\theta E\theta) + r\frac{\partial E\phi}{\partial\phi}\right].$$

Curl :

$$\vec{\nabla}.\vec{E} = \frac{1}{r^2\sin\theta}\begin{vmatrix} \hat{e}_r & r\hat{e}_\phi & r\sin\theta \hat{e}_\phi \\ \dfrac{\partial}{\partial r} & \dfrac{\partial}{\partial\phi} & \dfrac{\partial}{\partial\phi} \\ E_r & rE\theta & r\sin\theta E\phi \end{vmatrix}$$

Laplacian Operator :

$$\nabla^2 \times V = \frac{1}{r^2 \sin\theta}\left[\sin\theta\frac{\partial}{\partial r}(r^2\frac{\partial V}{\partial r}) + \frac{\partial}{\partial\theta}(\sin\theta\frac{\partial V}{\partial\theta}) + \frac{1}{\sin\theta}\frac{\partial^2 V}{\partial\phi^2}\right]$$

The space variables (x, y, z) in cartesian co-ordinates can be related to variables (r, θ, ϕ) of a spherical co-odinate system as;

$(x, y, z) \rightarrow (r, \theta, \phi)$

$$r = \sqrt{x^2 + y^2 + z^2}; \theta = \tan^{-1}\frac{\sqrt{x^2 + y^2}}{z}; \phi = \tan^{-1}\frac{y}{x}$$

or

$(r, \theta, \phi) \rightarrow (x, y, z)$

$x = r\sin\theta\cos\phi;\ \ y = r\sin\theta\sin\phi;\ \ z = r\cos\theta$

Unit vector transformation:

$\hat{er} = \sin\theta\cos\phi\,\hat{i} + \sin\theta\sin\phi\,\hat{j} + \cos\theta\,\hat{k}$

$\hat{e_\theta} = \cos\theta\cos\phi\,\hat{i} + \cos\theta\sin\phi\,\hat{j} - \sin\theta\,\hat{k}$

$\hat{e_\theta} = -\sin\phi\,\hat{i} + \cos\theta\,\hat{j}$

$\hat{i} = \sin\theta\cos\phi\,\hat{e_r} + \cos\theta\cos\phi\,\hat{e_\theta} - \sin\theta\,\hat{e_\phi}$

$\hat{j} = \sin\theta\cos\phi\,\hat{e_r} + \cos\theta\sin\phi\,\hat{e_\theta} + \cos\phi\,\hat{e_\phi}$

$\hat{k} = \cos\theta\,\hat{e_r} - \sin\theta\,\hat{e_\theta}$

Transformation of vectors :

Cartesian $(Ax, Ay, Az) \rightarrow$ cylindrical $(Ar, A\phi, Az)$.

$$\begin{pmatrix} A_r \\ A_\phi \\ A_z \end{pmatrix} = \begin{pmatrix} \cos\phi & \sin\phi & 0 \\ -\sin\phi & \cos\phi & 0 \\ 0 & 0 & 1 \end{pmatrix}\begin{pmatrix} A_x \\ A_y \\ A_z \end{pmatrix}$$

$\Rightarrow Ar = A_x\cos\phi + Ay\sin\phi$

$$A\phi = -A_x \sin\phi + Ay\cos\phi$$

$$Az = Az$$

$\therefore \vec{A} = \hat{i}\,Ax + \hat{j}\,Ay + \hat{k}\,Az$ in Cartesian co-ordinates; will be ;

$$\vec{A} = (Ax\cos\phi + Ay\sin\phi)\hat{er} + (-Ax\sin\phi + Ay\cos\phi)\hat{e\phi} + A_z\,\hat{ez}$$

Similarly; cylindrical $(Ar, A\phi, Az) \rightarrow$ Cartesian $(Ax, Ay, Az):-$

$$\begin{pmatrix} Ax \\ Ay \\ Az \end{pmatrix} = \begin{pmatrix} \cos\phi & -\sin\phi & 0 \\ \sin\phi & \cos\phi & 0 \\ 0 & 0 & 1 \end{pmatrix}\begin{pmatrix} Ar \\ A\phi \\ Az \end{pmatrix}$$

Cartesian $(Ax, Ay, Az) \rightarrow$ spherical $(A_r, A_\theta, A_\phi):-$

$$\begin{pmatrix} Ar \\ A\theta \\ A\phi \end{pmatrix} = \begin{pmatrix} \sin\theta\cos\phi & \sin\theta\sin\phi & \cos\phi \\ -\cos\theta\cos\phi & \cos\theta\sin\phi & -\sin\theta \\ -\sin\phi & \cos\phi & 0 \end{pmatrix}\begin{pmatrix} Ax \\ Ay \\ Az \end{pmatrix}$$

and spherical $(A_r, A_\theta, A_\phi) \rightarrow$ Cartesian (Ax, Ay, Az)

$$\begin{pmatrix} A_x \\ A_y \\ A_z \end{pmatrix} = \begin{pmatrix} \sin\theta\cos\phi & \cos\theta\cos\phi & -\sin\phi \\ \sin\theta\sin\phi & \cos\theta\sin\phi & \cos\phi \\ \cos\theta & -\sin\theta & 0 \end{pmatrix}\begin{pmatrix} A_r \\ A_\theta \\ A_\phi \end{pmatrix}$$

(P19) Two uniform vector fields are given by :

$$\vec{E} = -5\hat{e}_r + 10\,\hat{e\phi} + 3\,\hat{ez} \quad \& \quad \vec{F} = -\hat{e}_r + 2\,\hat{e\phi} - 6\,\hat{ez}, \text{ Find};$$

(a) $\left|\vec{E} \times \vec{F}\right|$ (b) the angle $\vec{E}$ makes with the surface z=3 at $P(5, \pi/2, 3)$

Solution :

Bath $\vec{E}\,\&\,\vec{F}$ are at cylindrical co-ordinate system.

(a) $\therefore \vec{E} \times \vec{F} = \begin{vmatrix} \hat{e}_r & \hat{e}_\phi & \hat{e}_z \\ -5 & 10 & 3 \\ 1 & 2 & -6 \end{vmatrix} = -66\,\hat{e}_r - 27\,\hat{e}_\phi - 20\,\hat{e}_z$

$$\therefore \left|\vec{E} \times \vec{F}\right| = (66^2 + 27^2 + 20^2)^{1/2} = 74.06$$

(b)

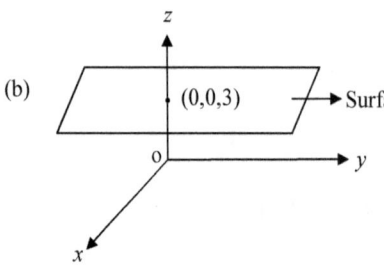

Surface at $z=3$

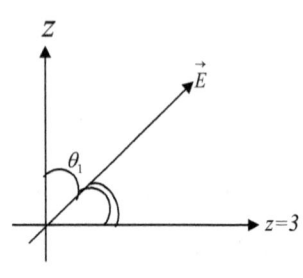

Utilizing the fact that the z-axis is normal to the surface $z=3$, the angle between the z-axis and $\vec{E}$ as shown in the above fig.

$$\vec{E} \cdot \hat{e}_z = |E||1|\cos\theta_1 \Rightarrow 3 = \sqrt{134}\cos\theta_1$$

$$\therefore \theta_1 = \cos^{-1}\frac{3}{\sqrt{134}} = 0.2592 = 74.98°$$

$\therefore$ The angle between the $z=3$ and $\vec{E}$ is ; $90° - \theta_1 = 15.02°$

(P20) Given a vector field; $\vec{D} = r\sin\theta\,\hat{e}_r - \dfrac{1}{r}\sin\theta\cos\phi\,\hat{e\theta} + r^2\,\hat{e\theta} + r^2\,\hat{e\theta}$

Determine $\vec{D}$ at $P(10,150°,300°)$.

Solution :

$$\vec{D} = 10\sin330°\,\hat{er} - \frac{1}{10}\sin150°\cos330°\,\hat{e\theta} + 100\,\hat{e\theta}$$

or, $\vec{D} = -5\,\hat{er} - 0.043\,\hat{e\theta} + 100\,\hat{e\theta}$.